COPYRIGHT INFORMATION

TITLE: Sea Creatures of the Sea & the Ocean

FIRST EDITION

Paperback Edition

Copyright © 2023 by: Michael Eric Nelson

Publishing by Padrí Miquel Publishing
www.elPadri.com

ISBN: 9798396351134

<u>DEDICATION</u> - <u>Dedicació</u>

Creatures of the Sea & Ocean with the Mazes included I dedicate to: first la princessa Sara amb Lola then my namesake Princep Eric, and to Arnau & la princessa Lola hermano Oriol the Great, and of course George the Curious and my 'lil man TEO, also to Kennedy and Melody & though at the time of authoring he cannot walk, to my strong man IVAR, I truly hope each of you have an opportunity to visit Rhode Island, experiencing the Atlantic Ocean as I did when I was a child, all the creatures and fish and fun in the sand (del mar) and fishing.

I Love each and everyone of you unconditionally, as this is what Love means. You each are great, wise and important. You should remember the greatest of commandments to always LOVE thy neighbor as thyself, for with this one act you shall always be on the right path, doing all you do in the name of LOVE.

<u>Pero REMEMBER:</u>

Joc Soc el Padrí sus plau Respect'm, Nens amb Nenas!

Criaturas del mar y el oceano
Creatures of the Sea and the Ocean
Foreword - Preface - Prefacio

This book is packed with fun and exciting coloring pages revolving around everything the sea and the ocean.   This is a wonderful opportunity to learn some English (ingles) words about the sea and about the ocean. This is also a great book filled with MAZES, and these mazes help develop hand eye coordination and we have even included all the answers/resolutions of the mazes in the back (no cheating). According to the NOAA In terms of geography, seas are smaller than oceans and are usually located where the land and ocean meet. An example of a "sea" is the Mediterranean "sea" vs. the Atlantic Ocean which the Mediterranean sea feeds into through the strait of Gibraltar, between the Pillars of Hercules, which the two pillars on the Spanish Flag and Coat of Arms represent. Typically, seas are partially enclosed by land. Seas are found on the margins of the ocean and are partially enclosed by land.

Saltwater Fish Species (Sharks, Rays, Mackerel, Snappers, Flounder, Barracuda, Moray Eels, Tuna, Marlin, etc.)

What's the most common fish in the ocean? The most common fish is any of the species of a deepwater fish sometimes called a "bristle mouth." The fish is about the size of a small minnow. It is caught at 500 meters or deeper all over the world.

Coloring Pages

CLOWN FISH

YELLOW FISH

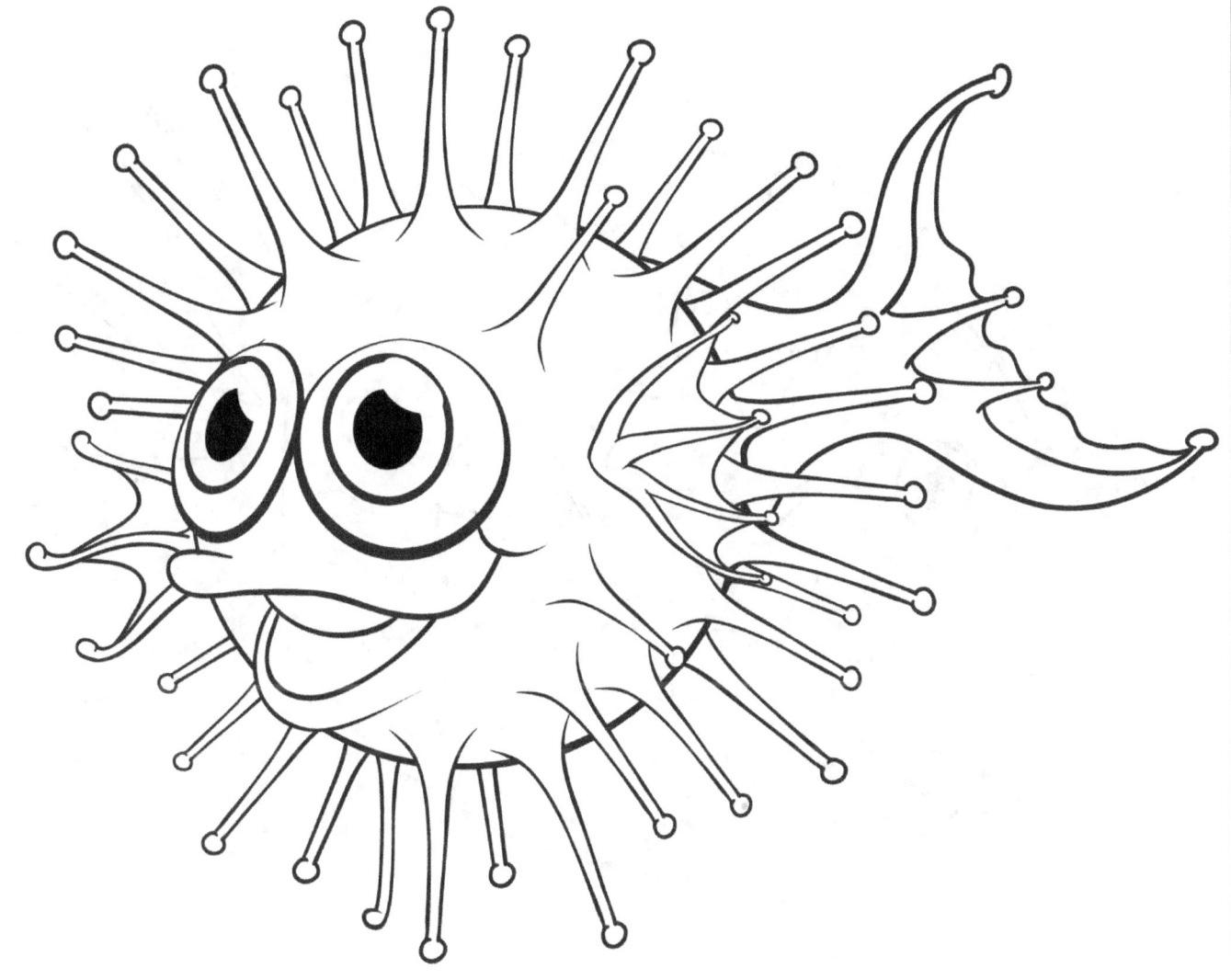

PUFFER FISH

BUBBLE FISH

PIRANHA

TINY FISH

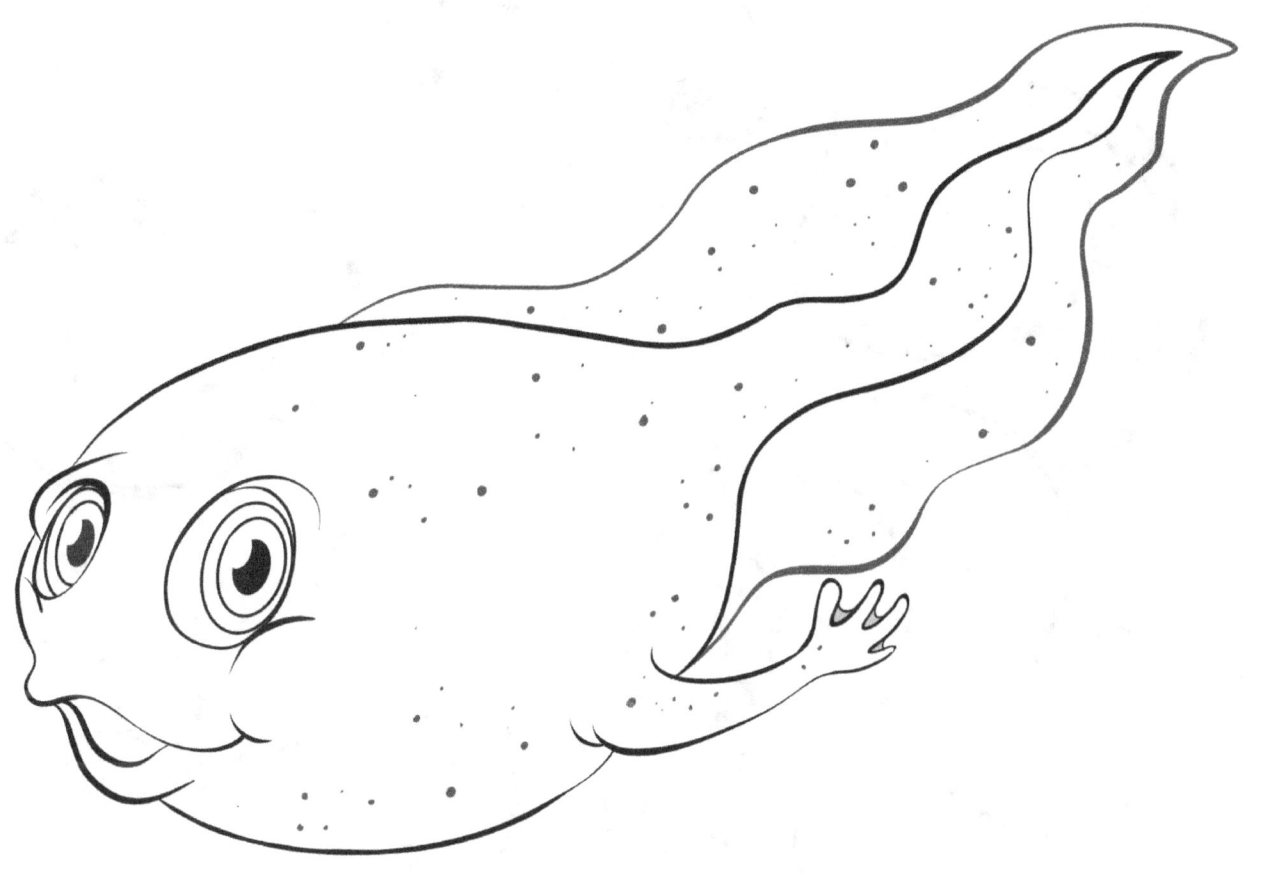

TADPOLE

OCTOPUS

OYSTER

XIPHIAS

STAR FISH

JELLY FISH

PUFFER FISH

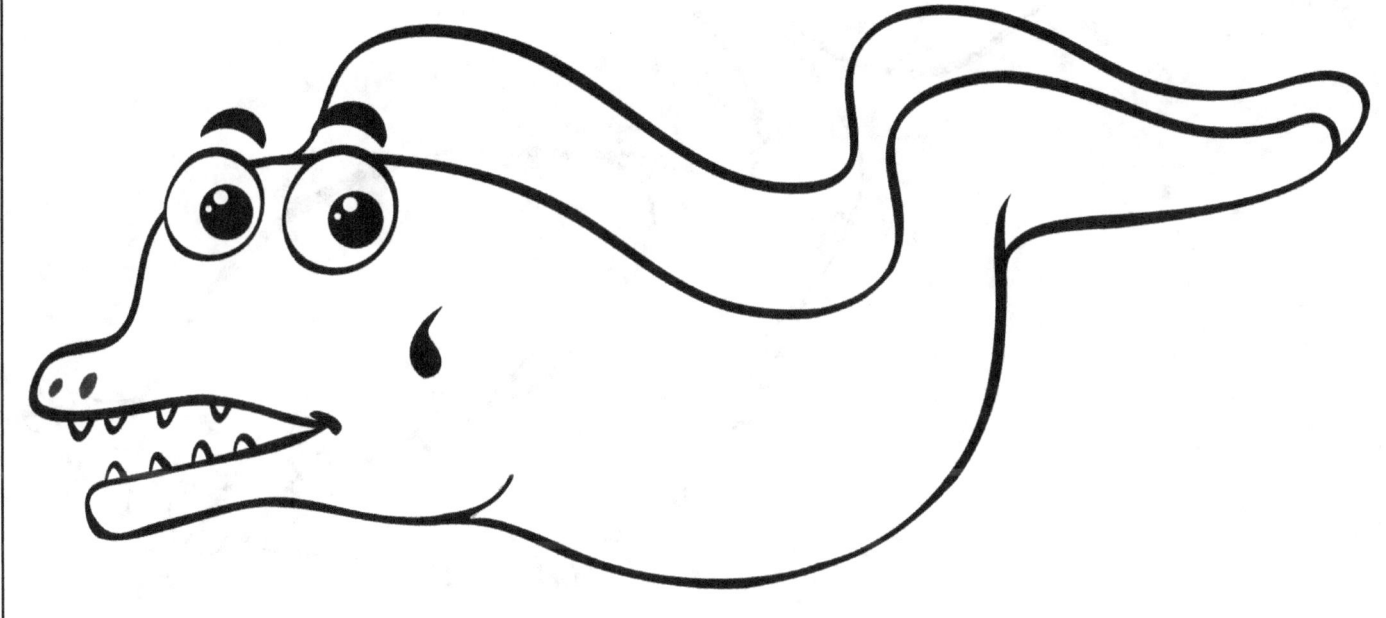

EEL

PELICAN

HAMMERHEAD SHARK

RAY

SEA URCHINE

WHALE

SHRIMP

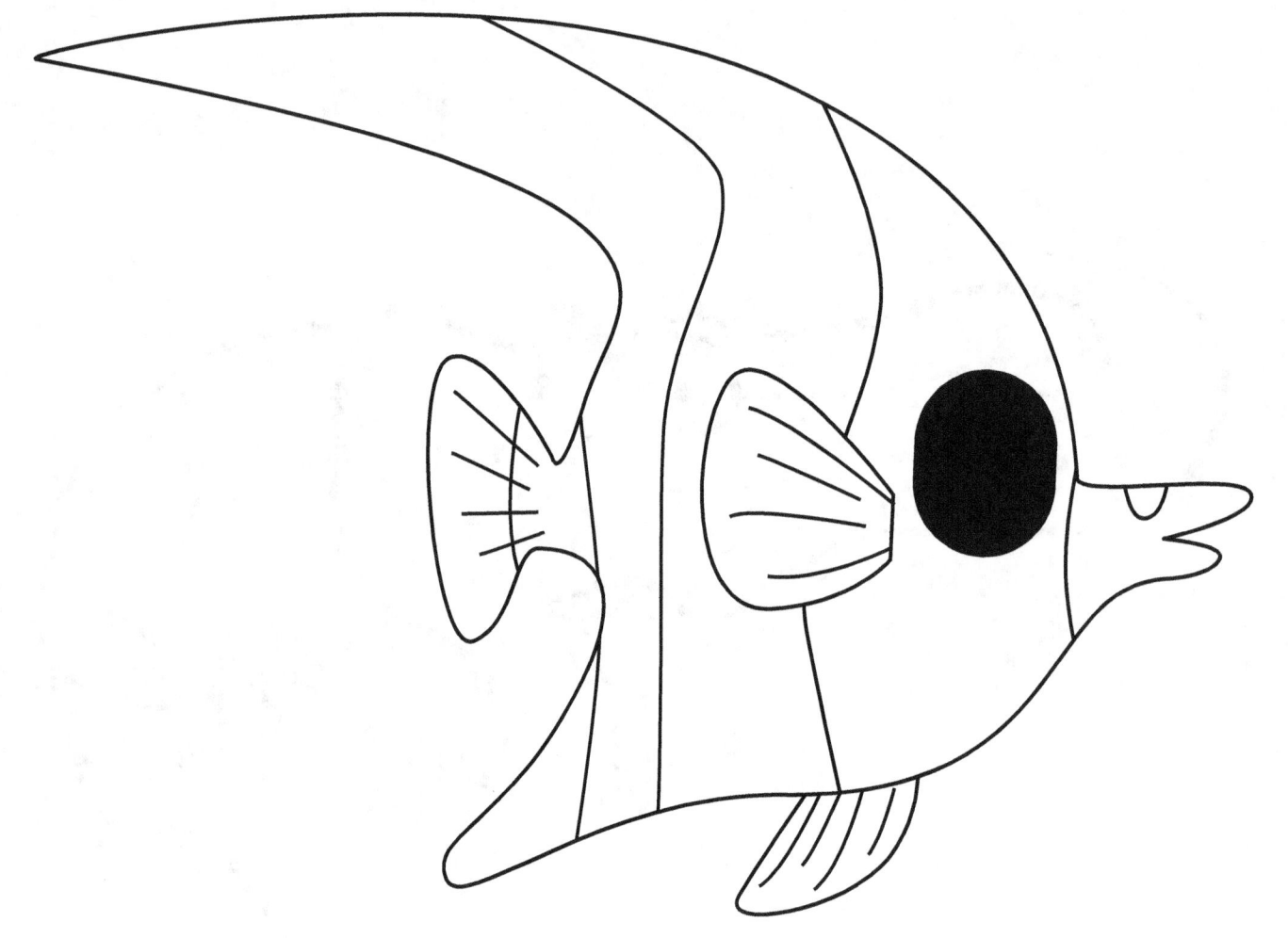

ANGEL FISH

CLOWNFISH

SEAL

DOLPHIN

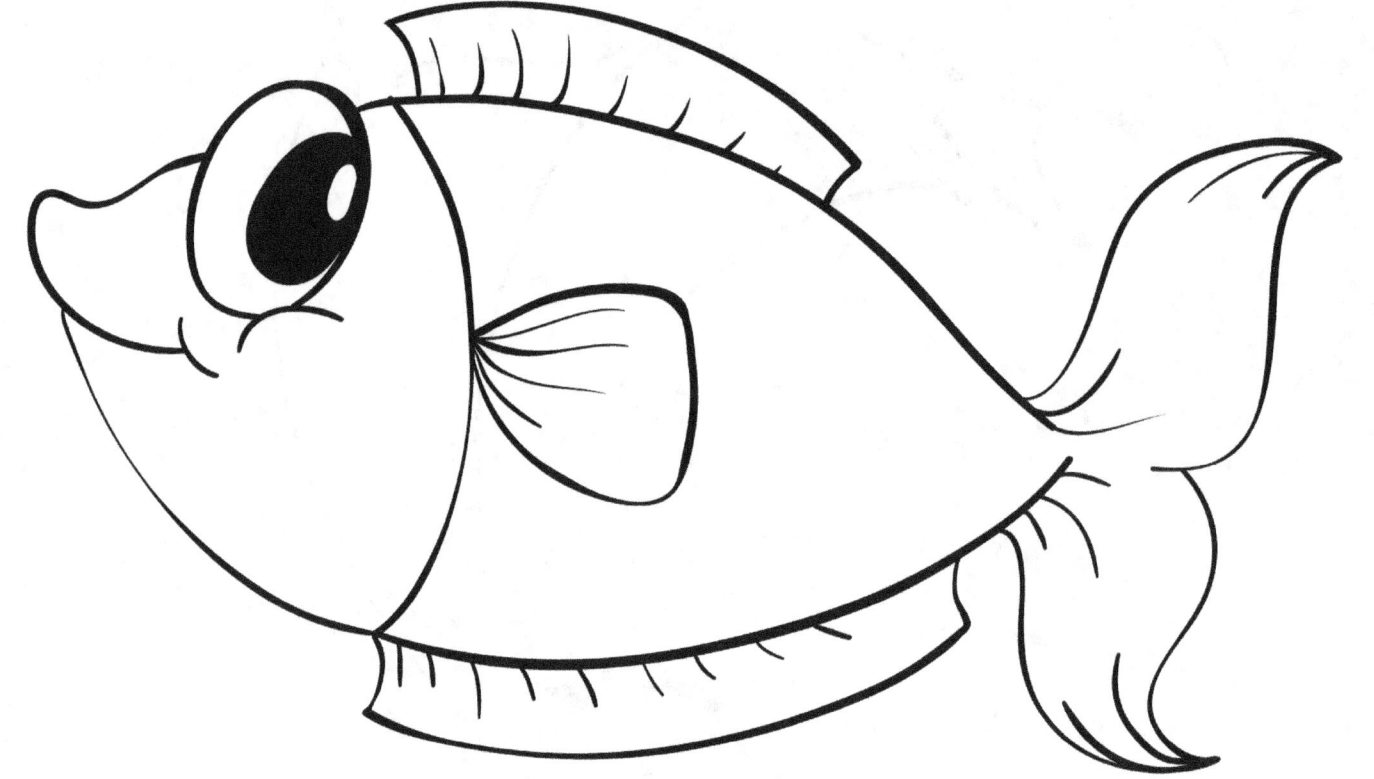

FISH

SEAHORSE

LOBSTER

SEA TURTLE

CRAB

SHARK

GOLD FISH

SQUID

WHALE

Mazes

MAZE - 1

MAZE - 2

MAZE - 3

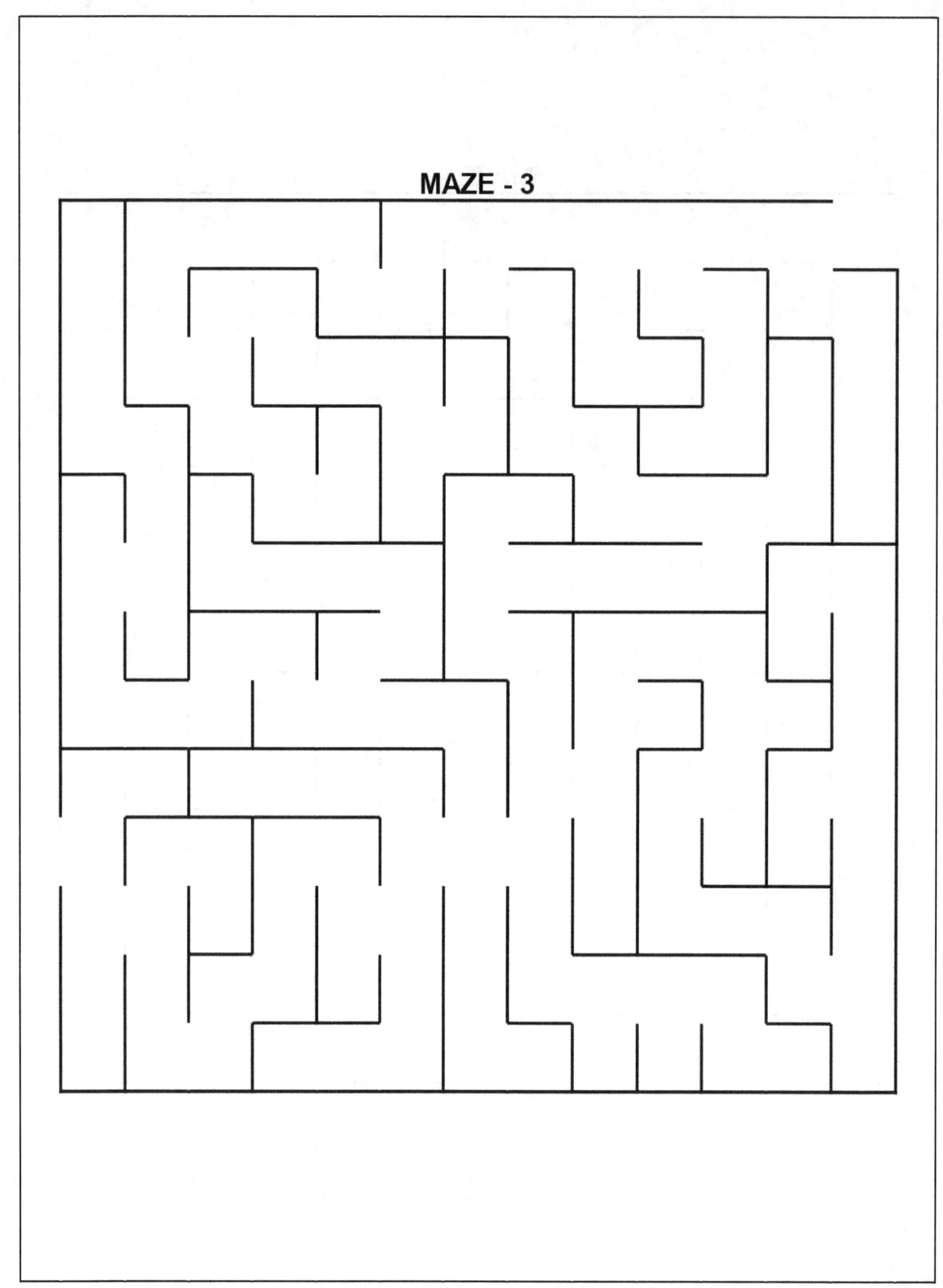

MAZE - 4

MAZE - 5

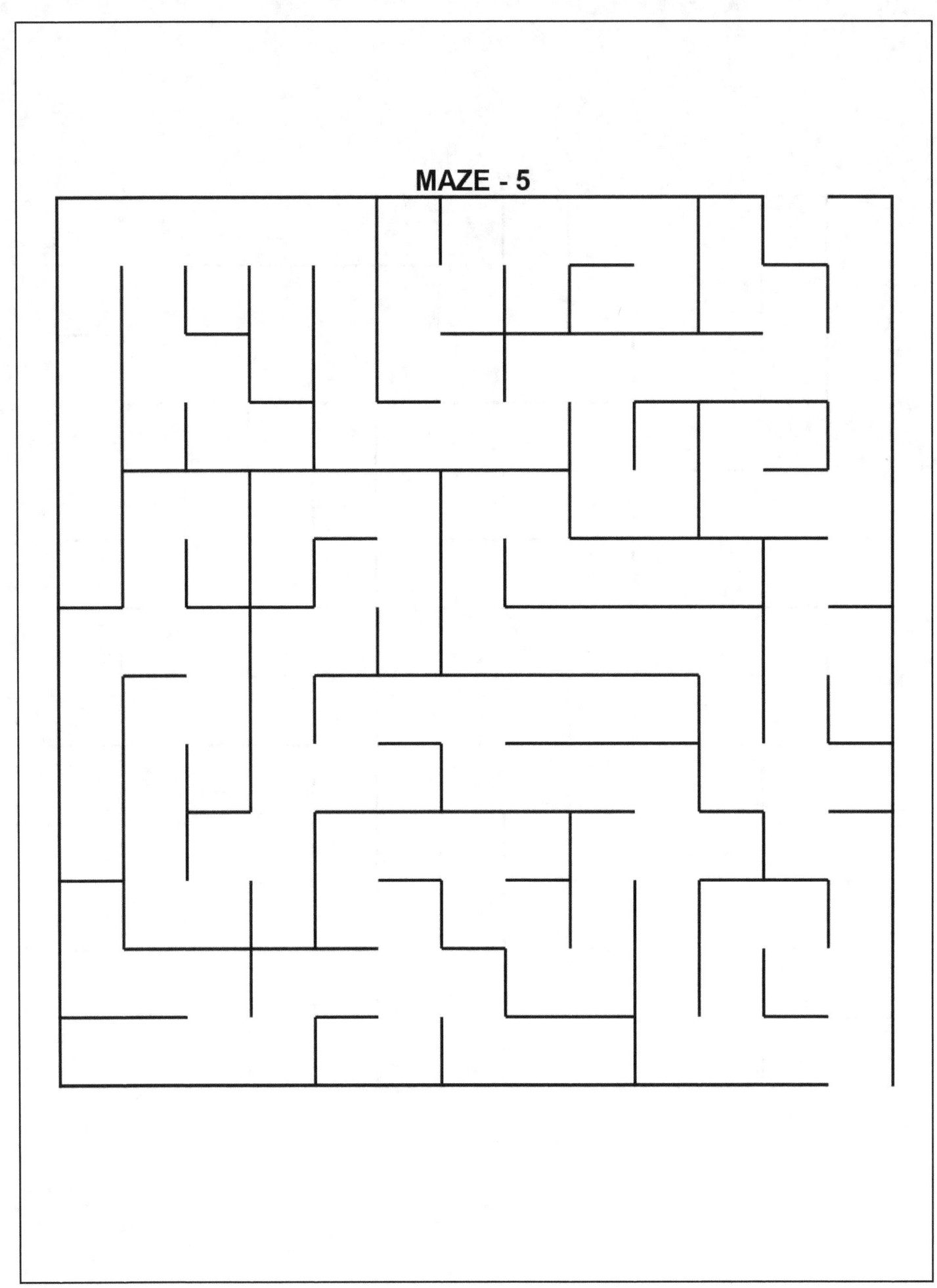

MAZE - 6

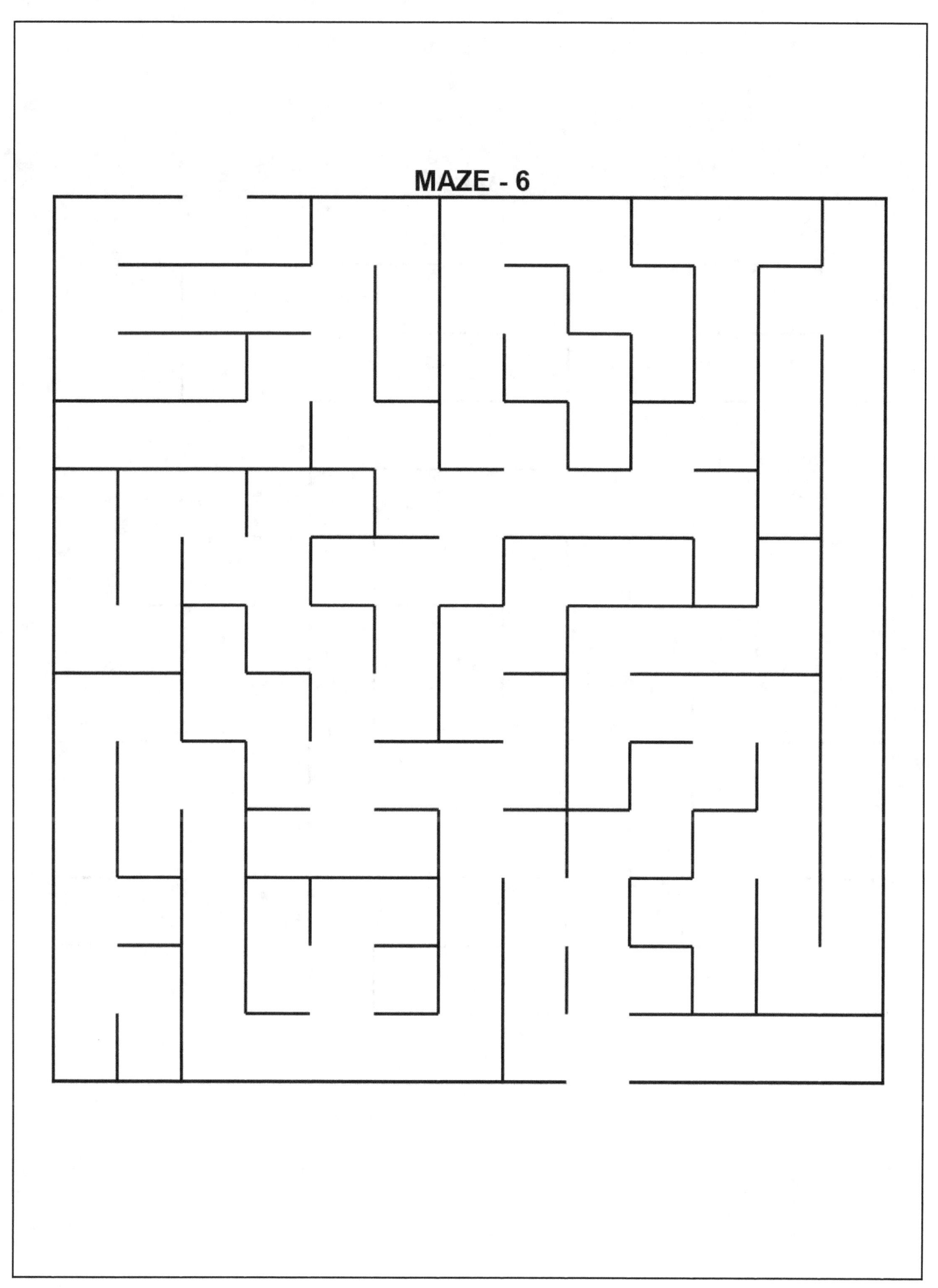

MAZE - 7

MAZE - 8

MAZE - 9

MAZE - 10

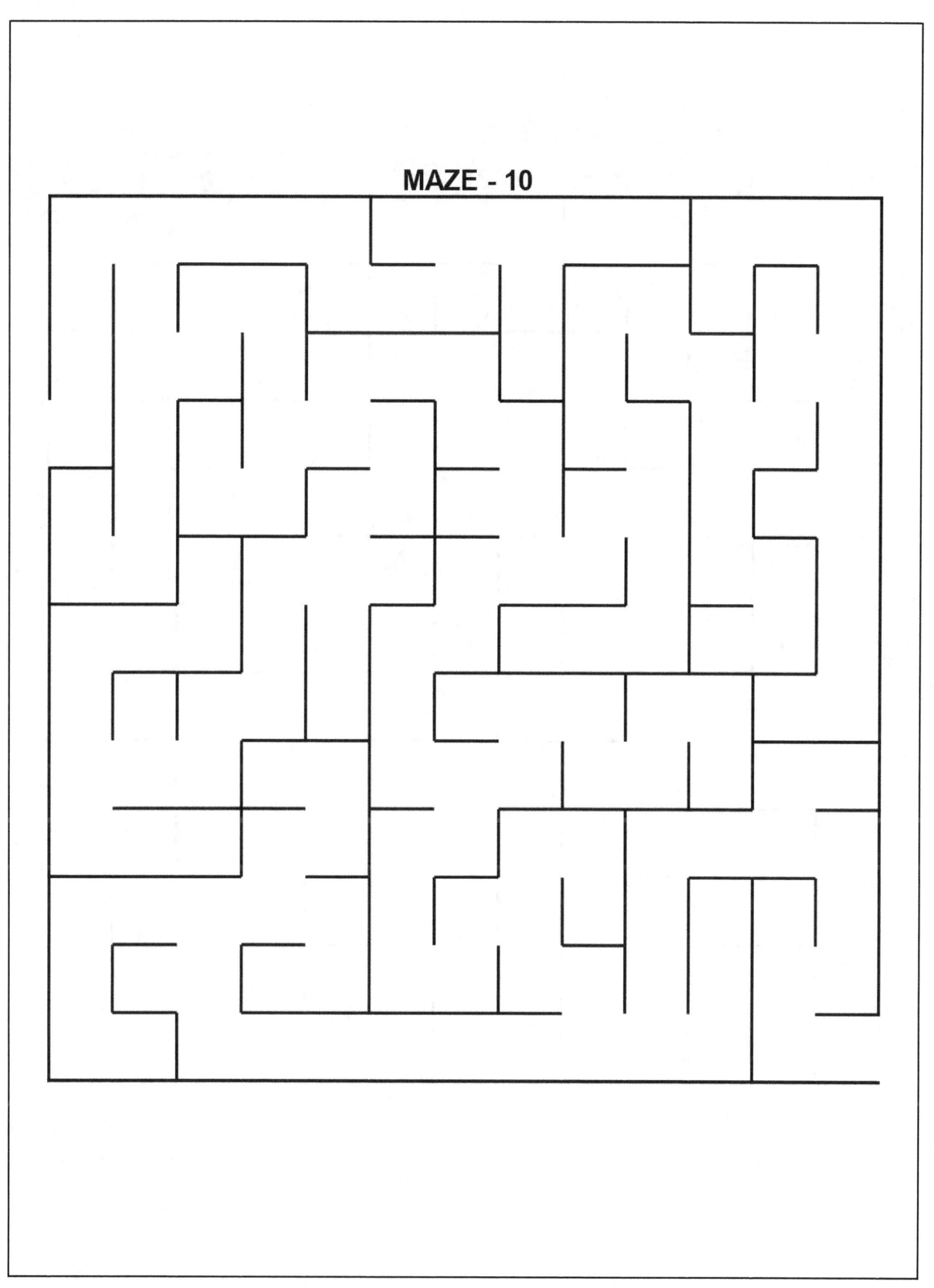

MAZE - 11

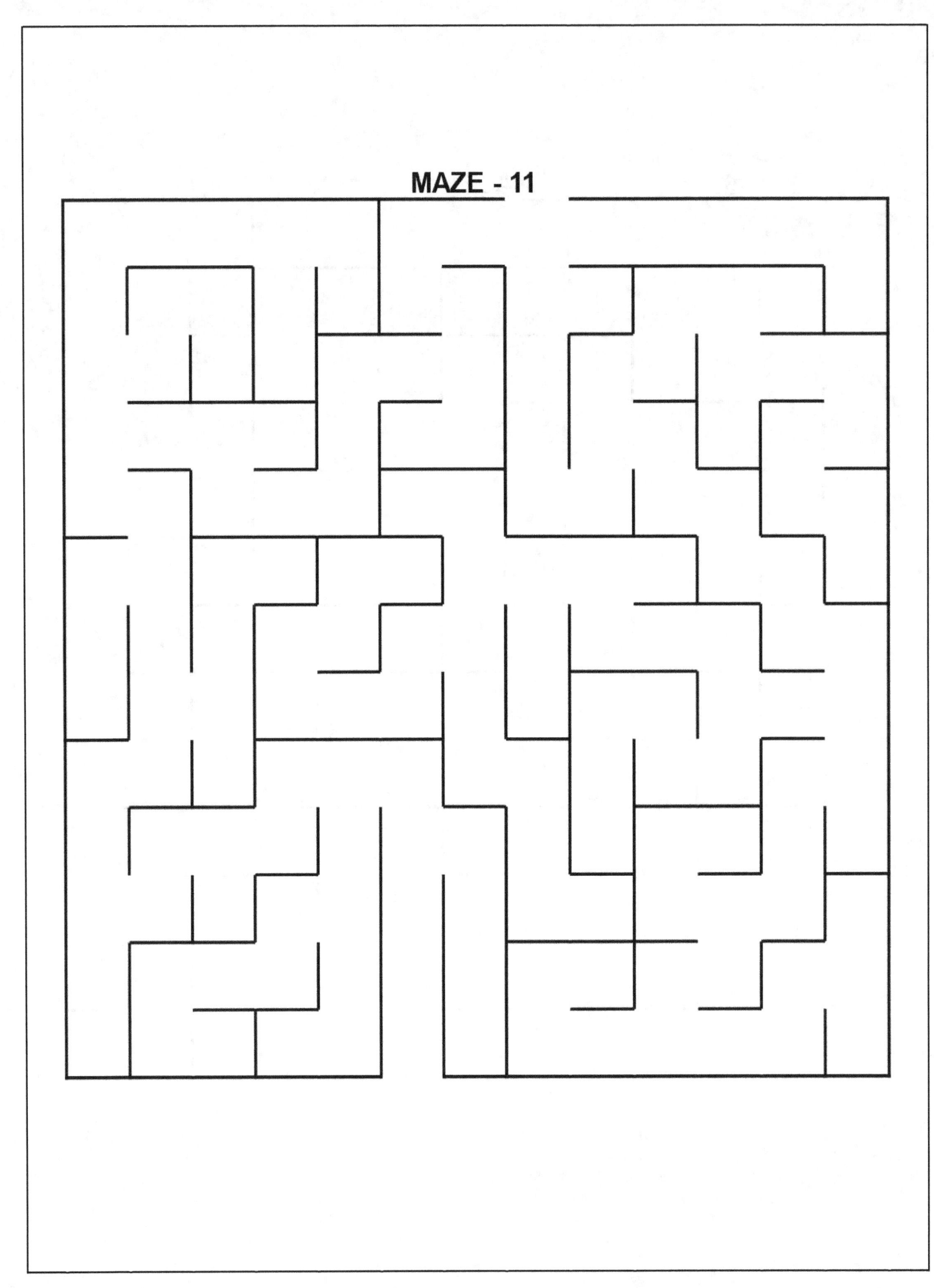

MAZE - 12

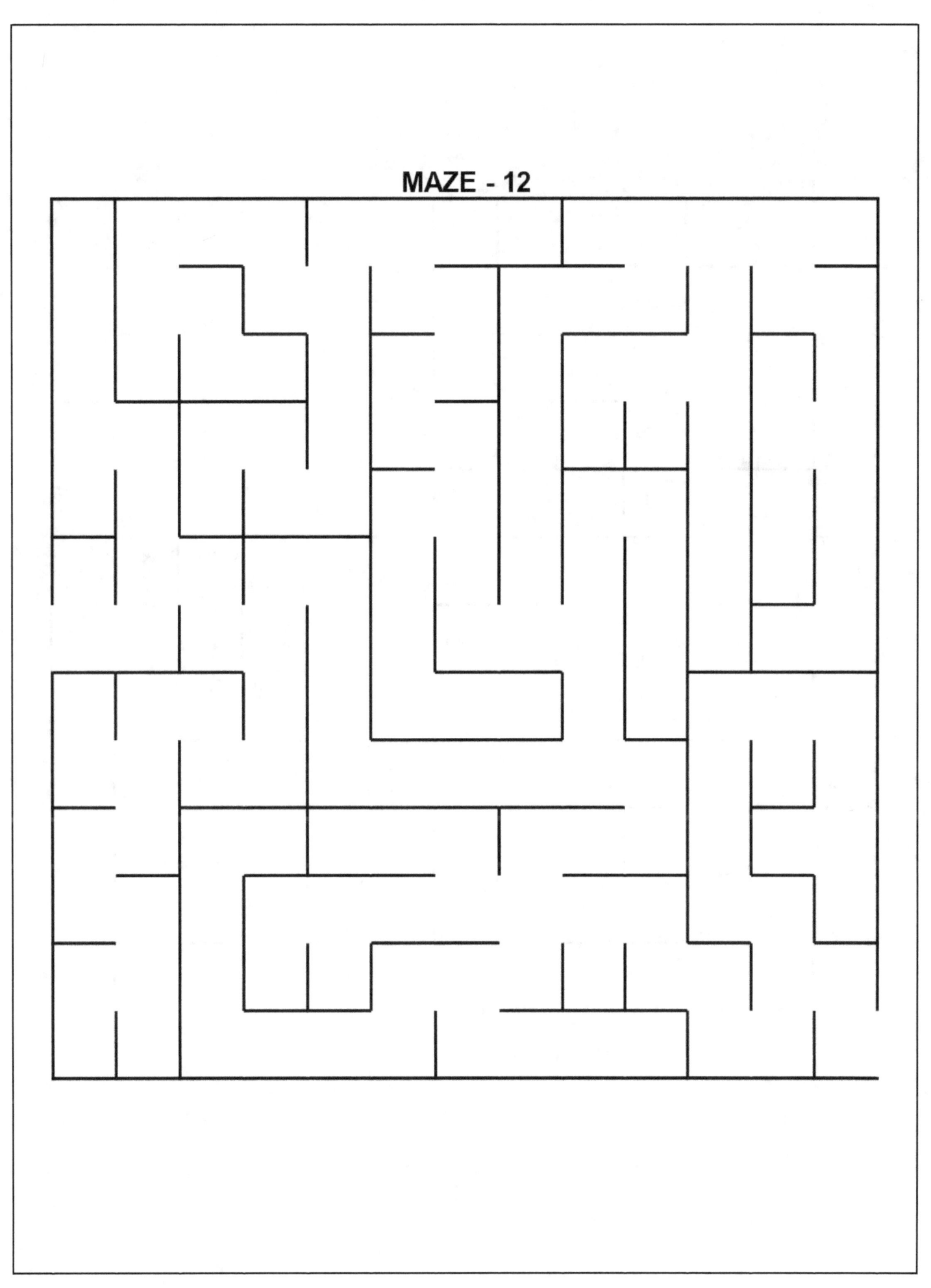

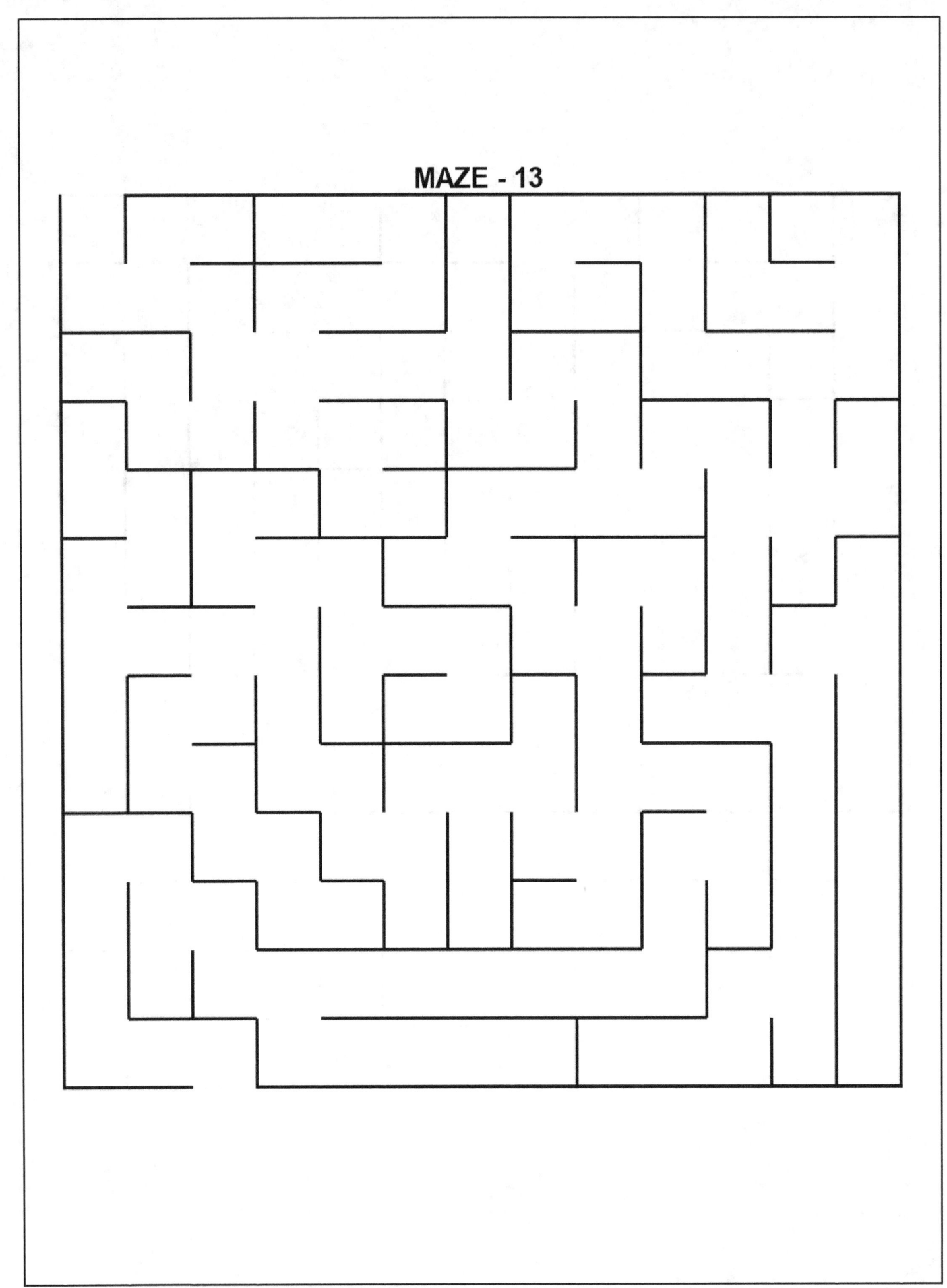

MAZE - 14

MAZE - 15

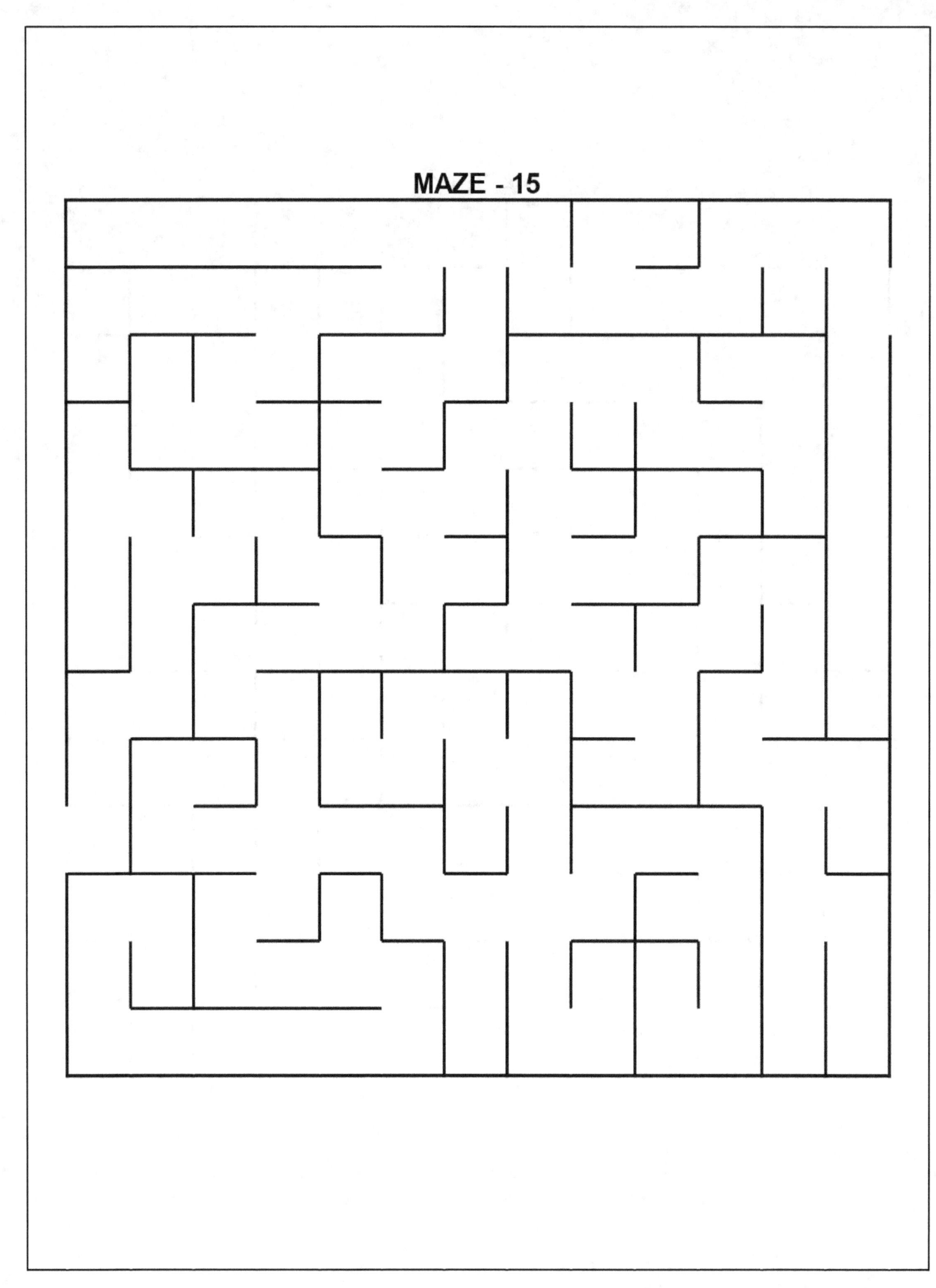

MAZE - 16

MAZE - 17

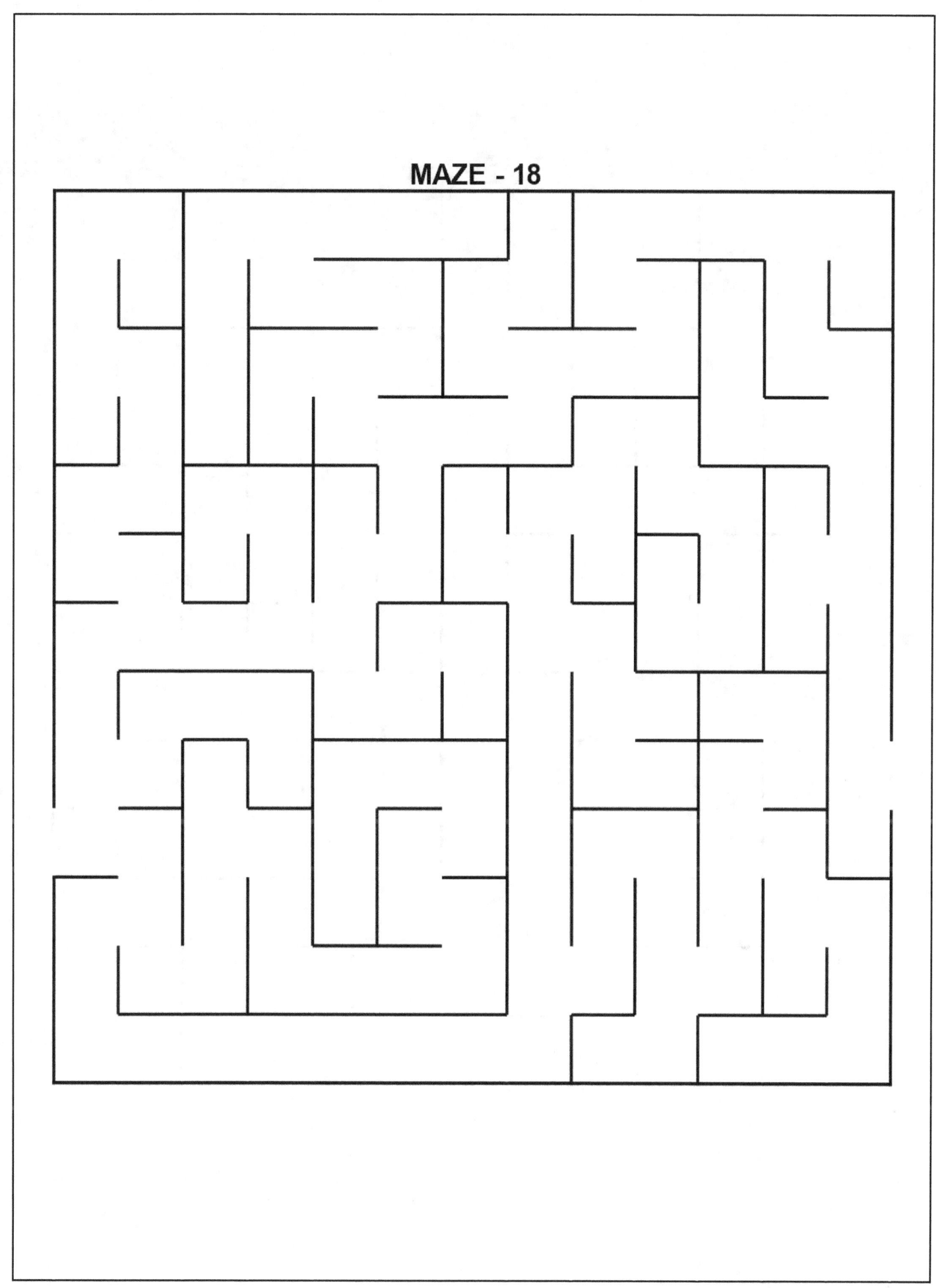

MAZE - 19

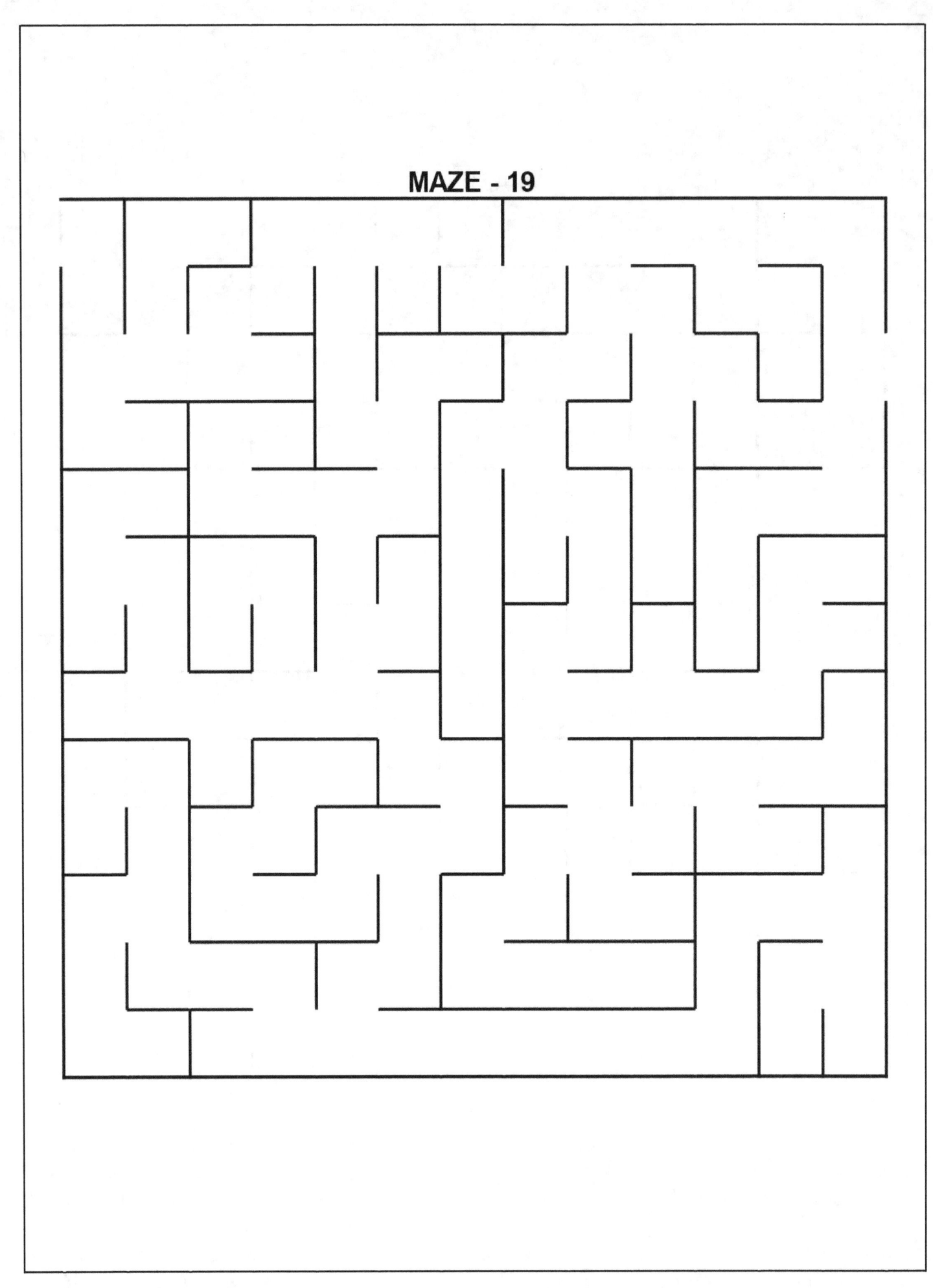

MAZE - 20

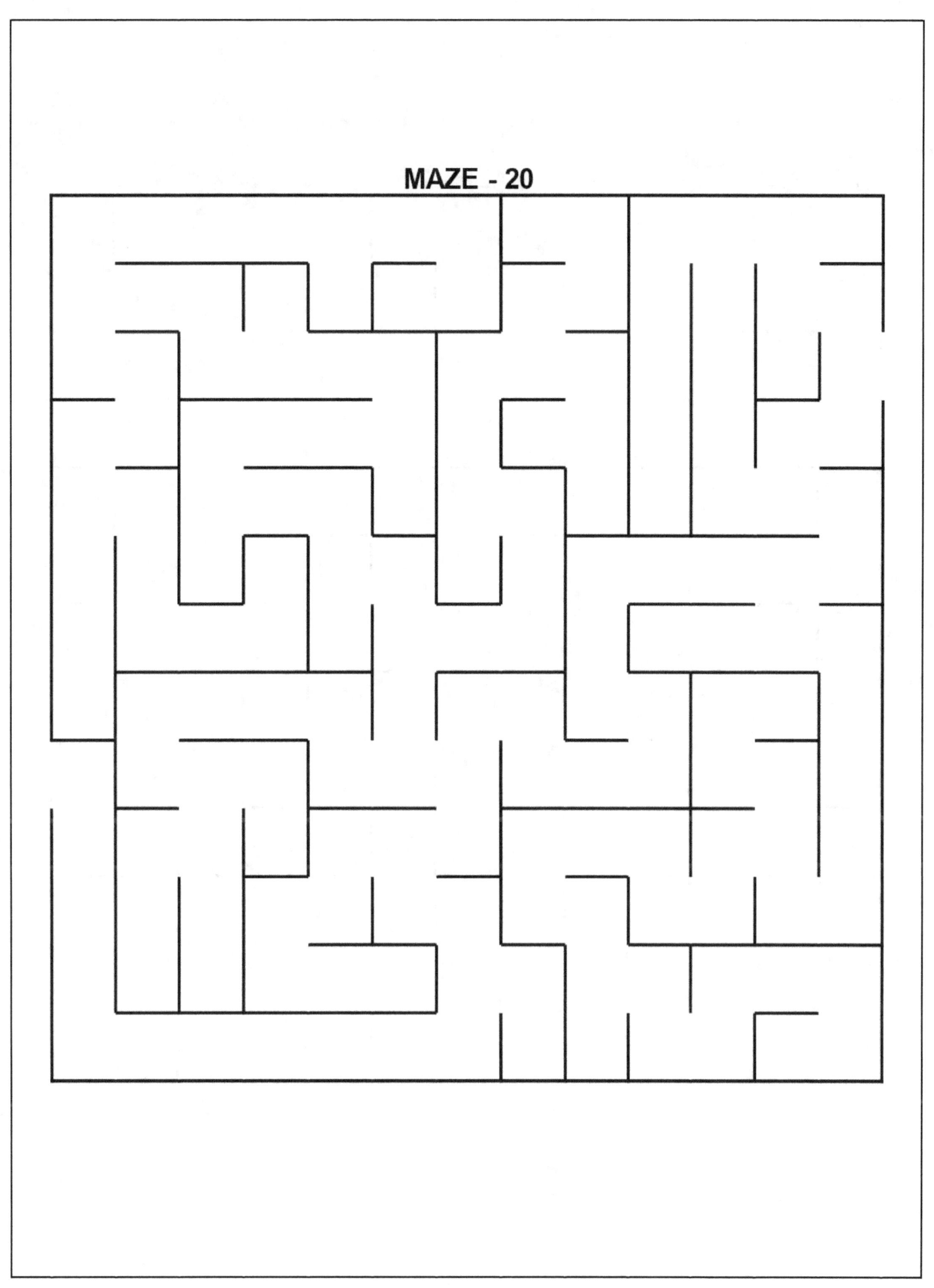

MAZE - 21

MAZE - 22

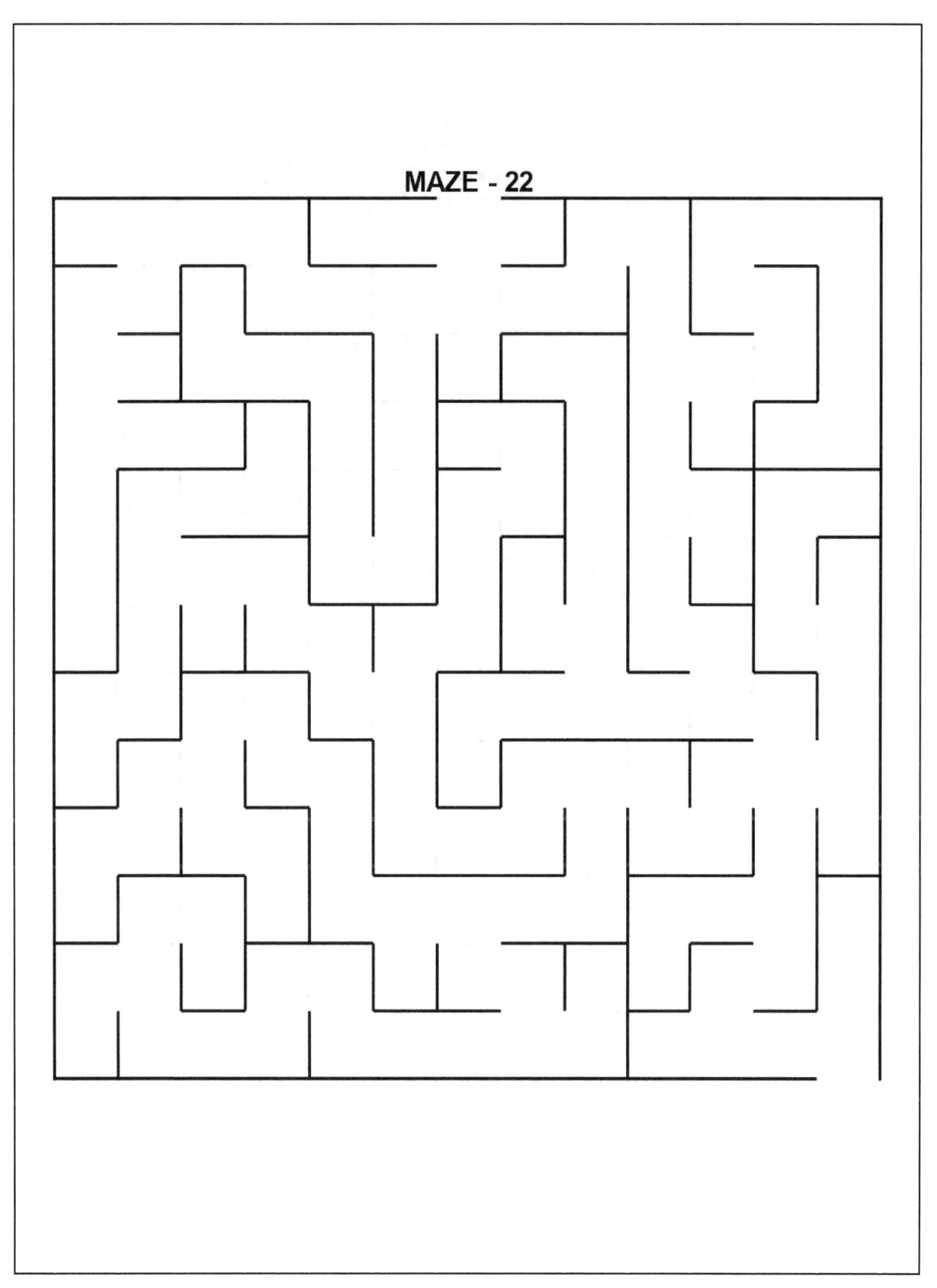

MAZE - 23

MAZE - 24

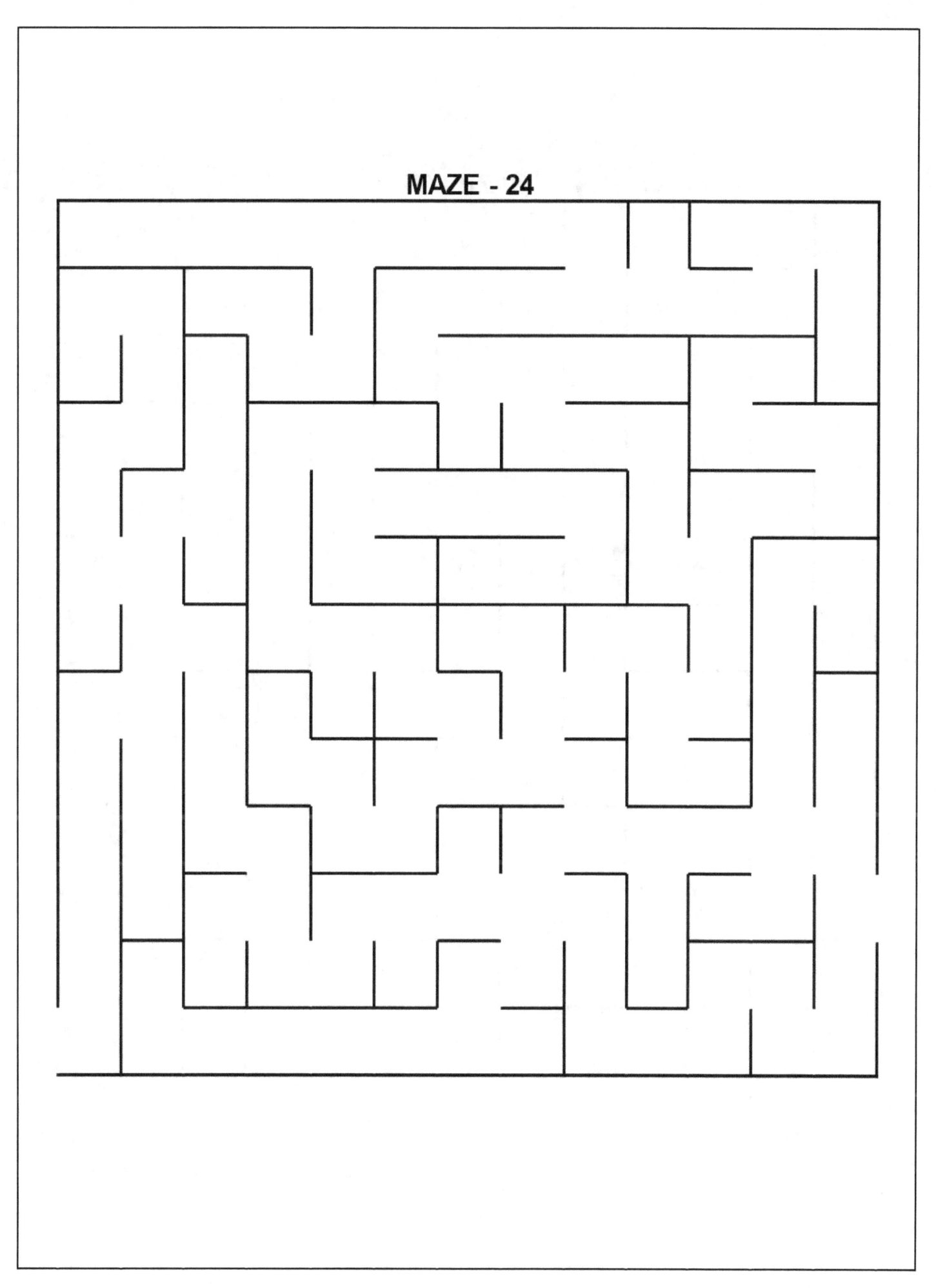

MAZE - 25

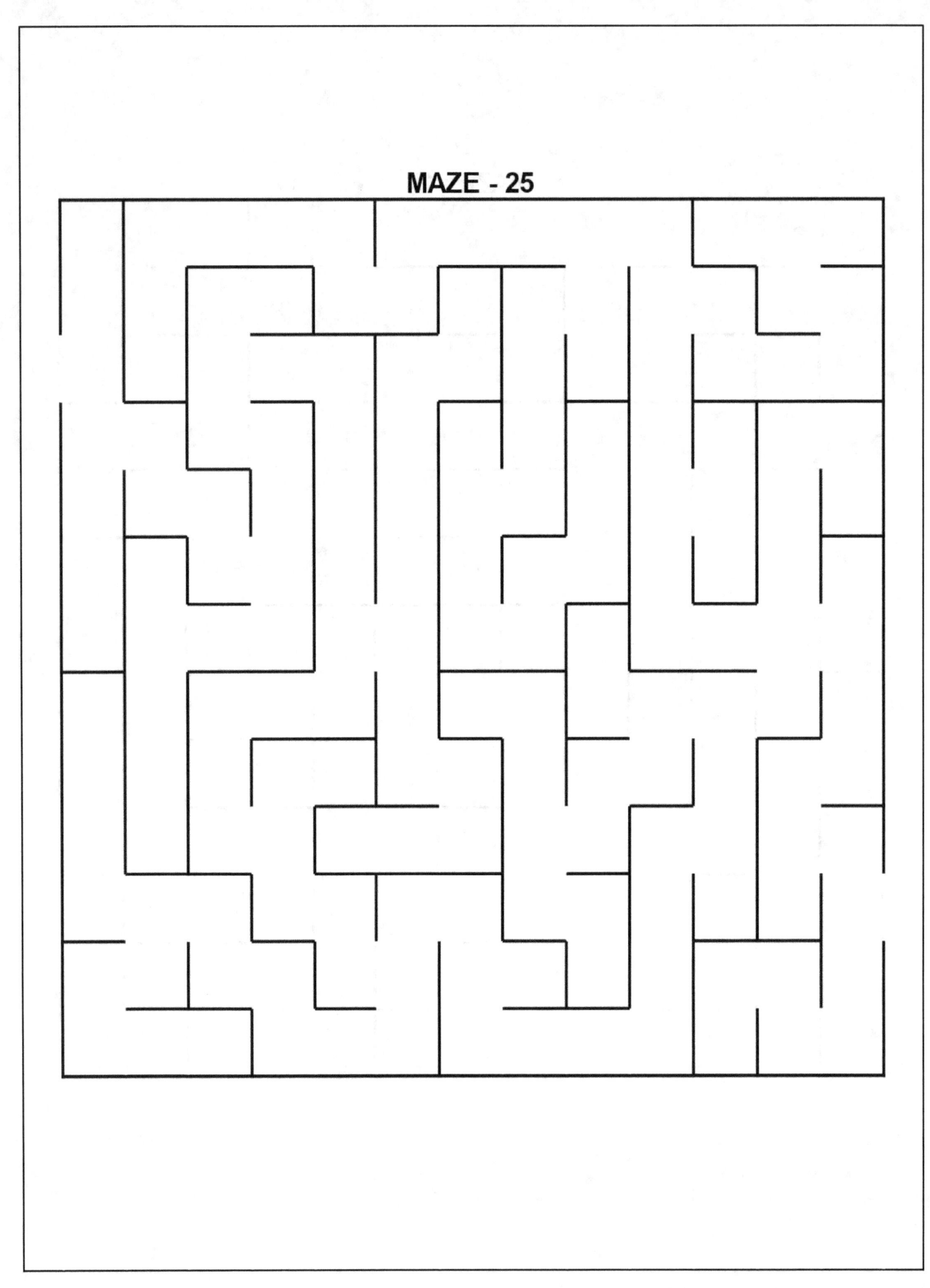

MAZE - 27

MAZE - 28

MAZE - 30

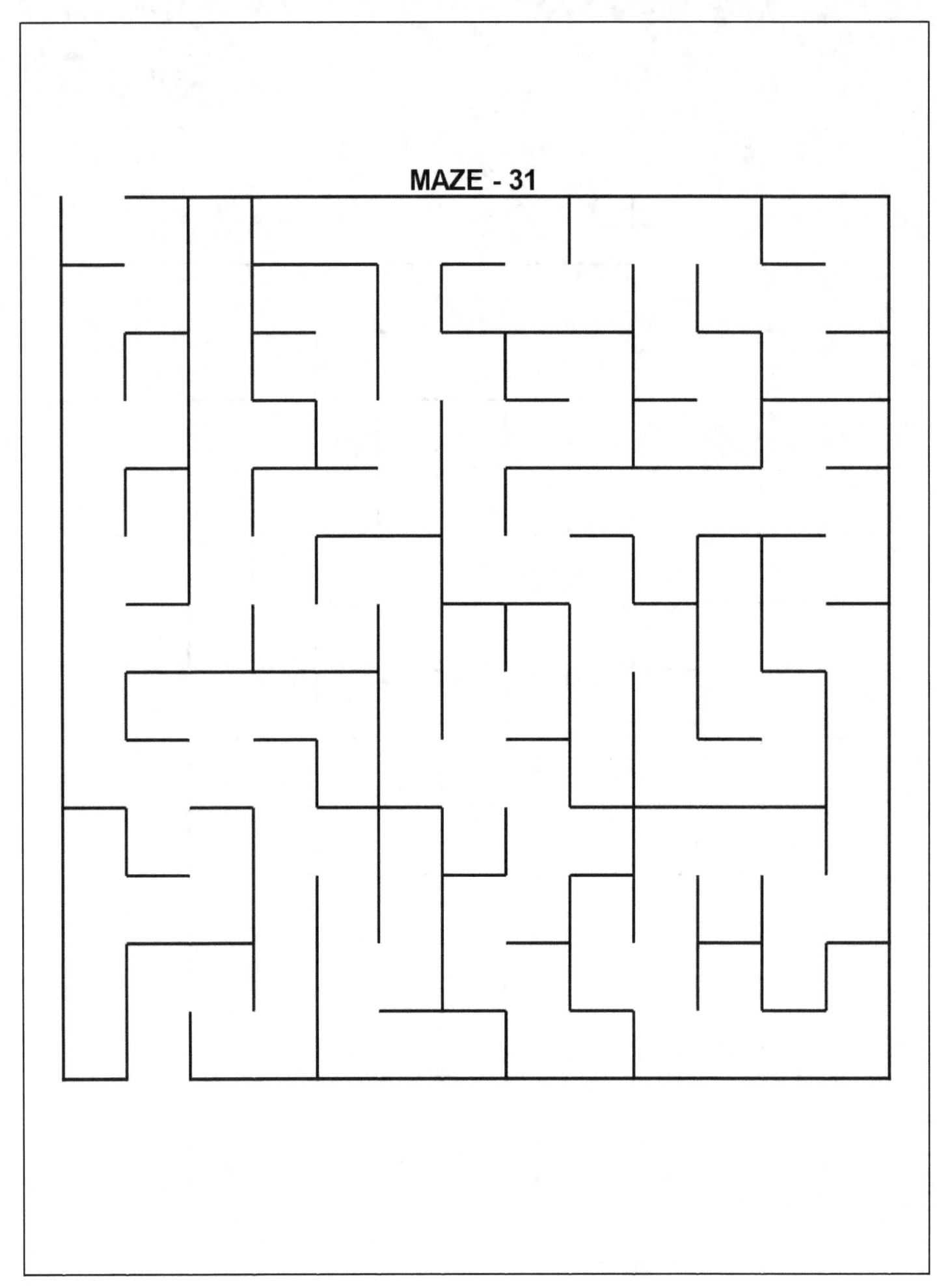

MAZE - 32

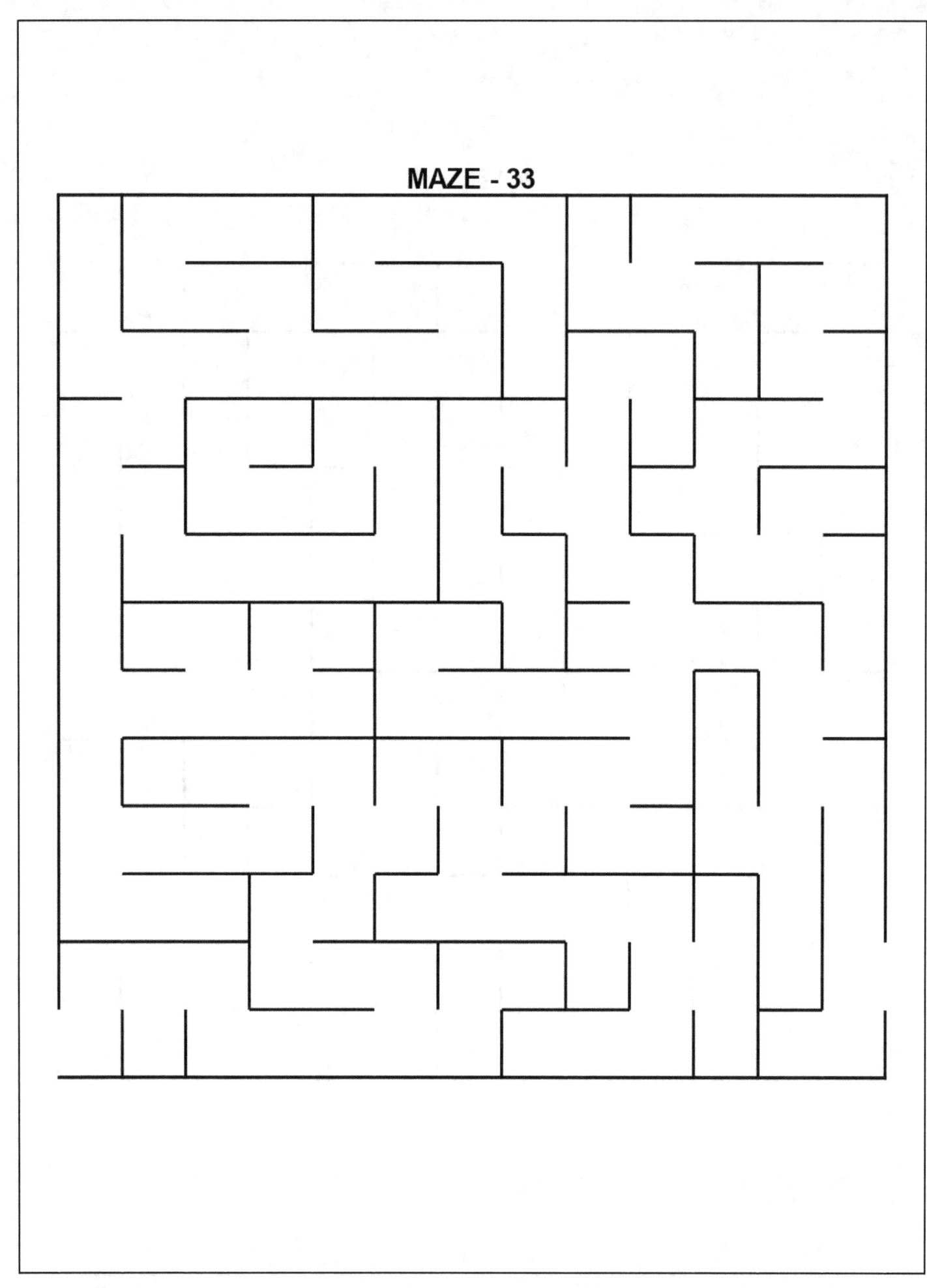

MAZE - 34

MAZE - 35

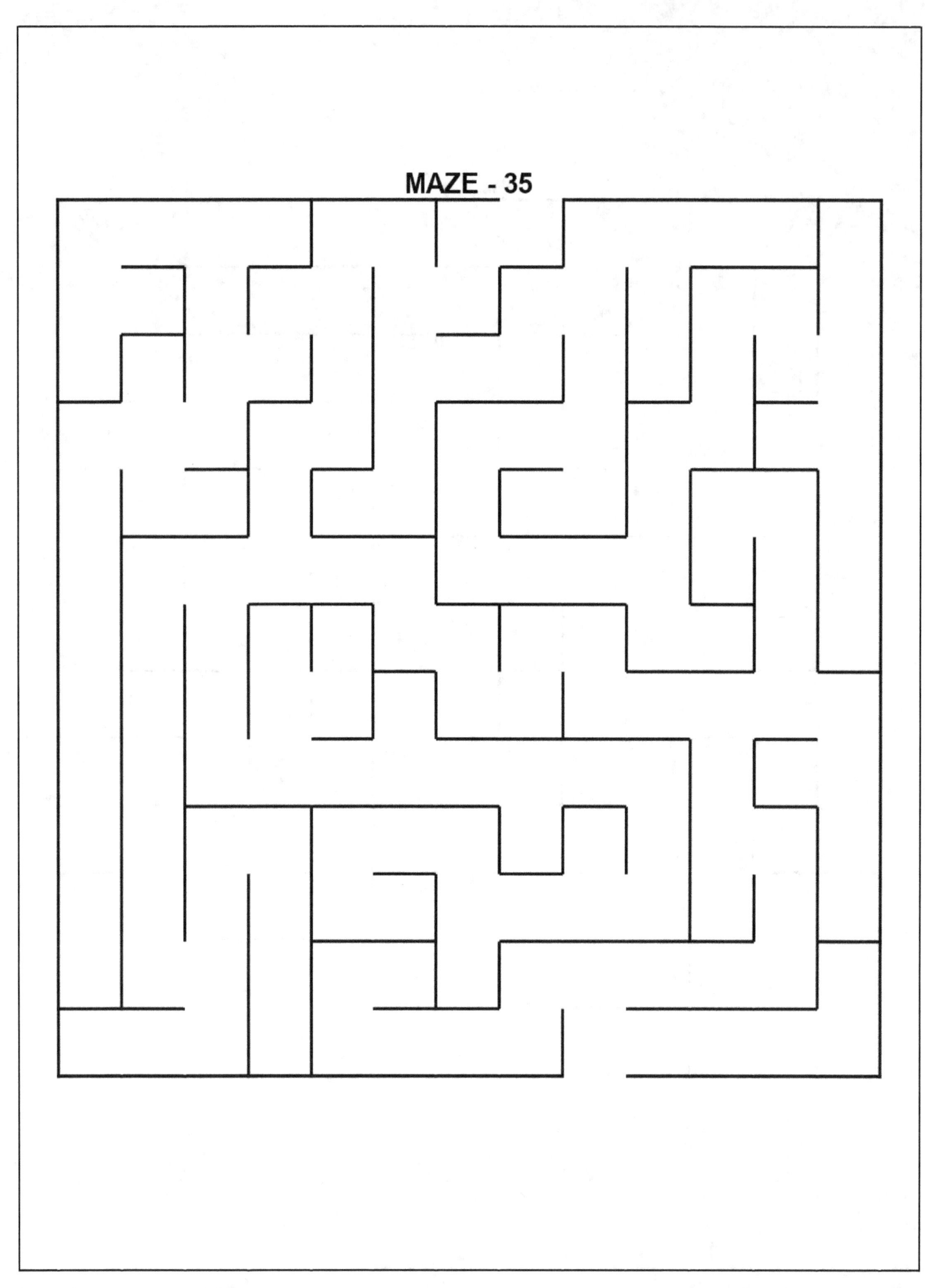

Mazes Solution

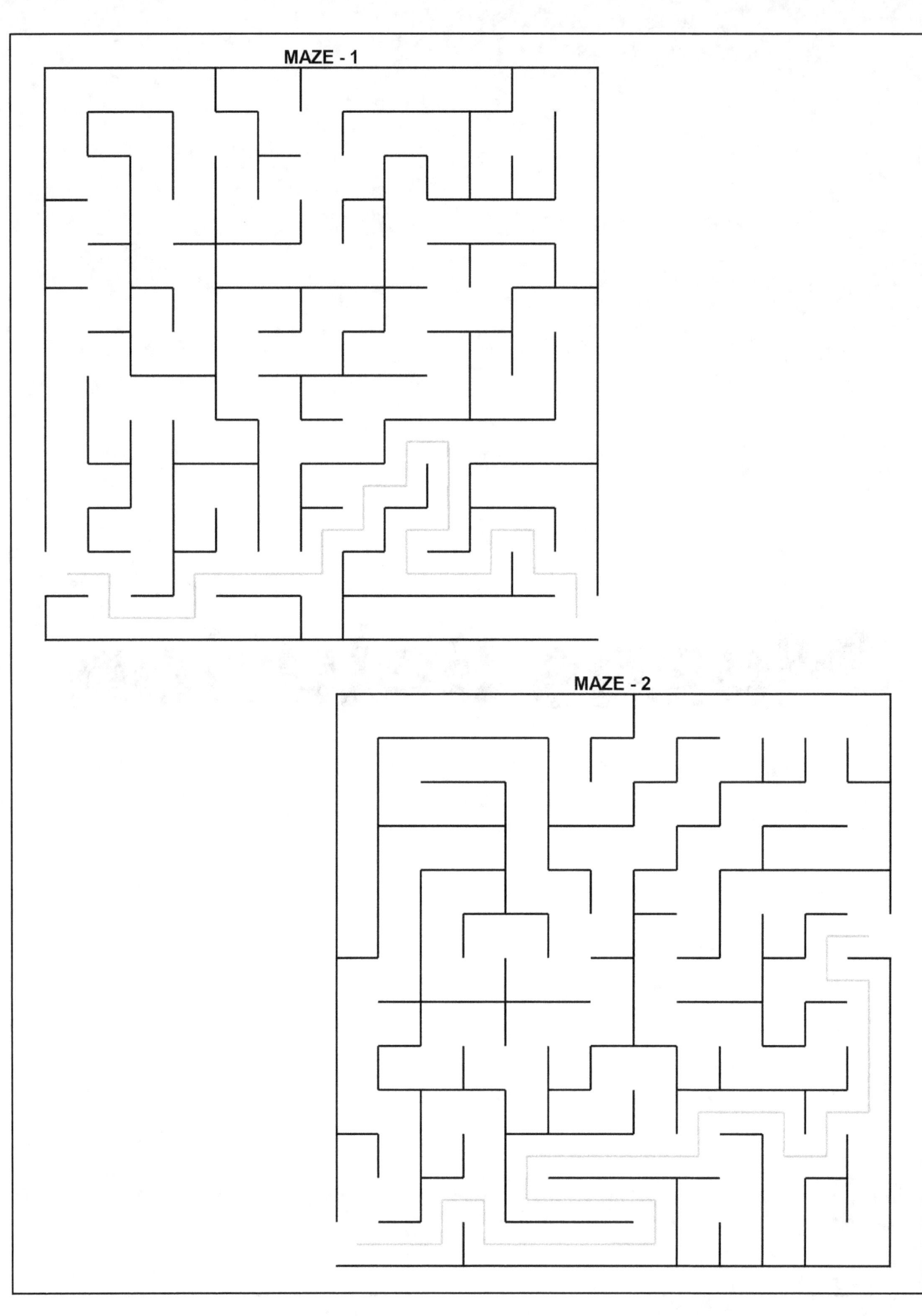

MAZE - 1

MAZE - 2

MAZE - 3

MAZE - 4

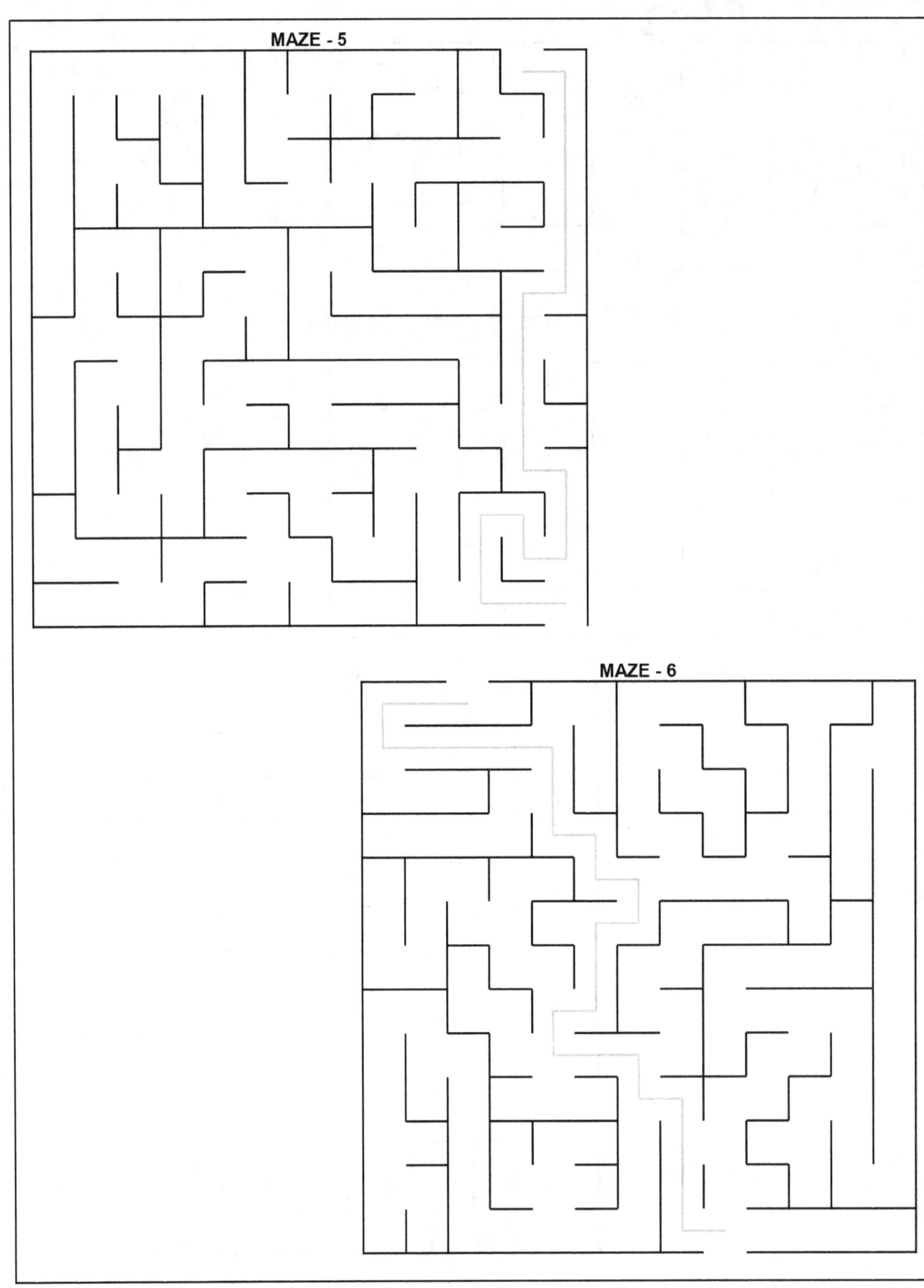

MAZE - 5

MAZE - 6

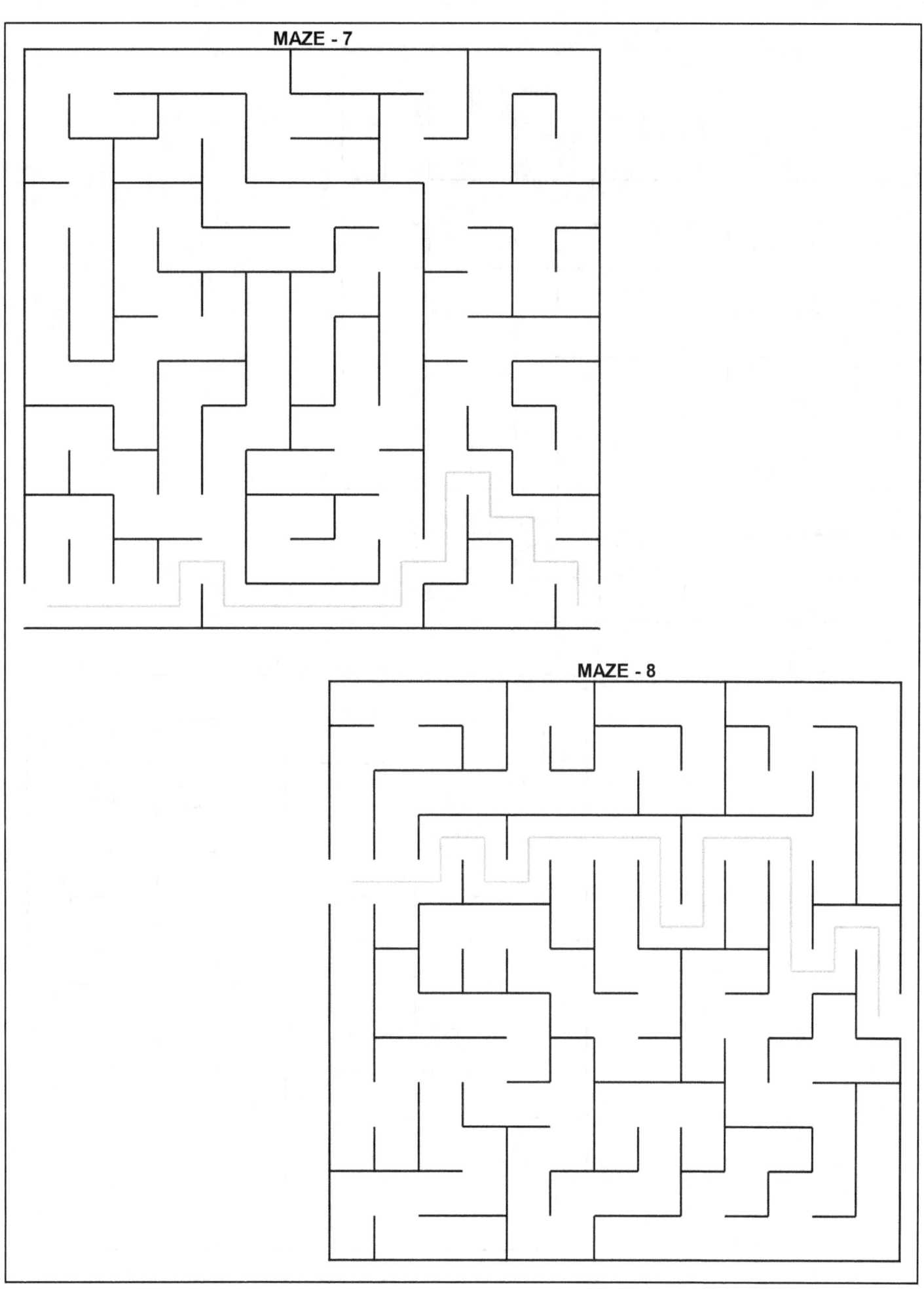

MAZE - 7

MAZE - 8

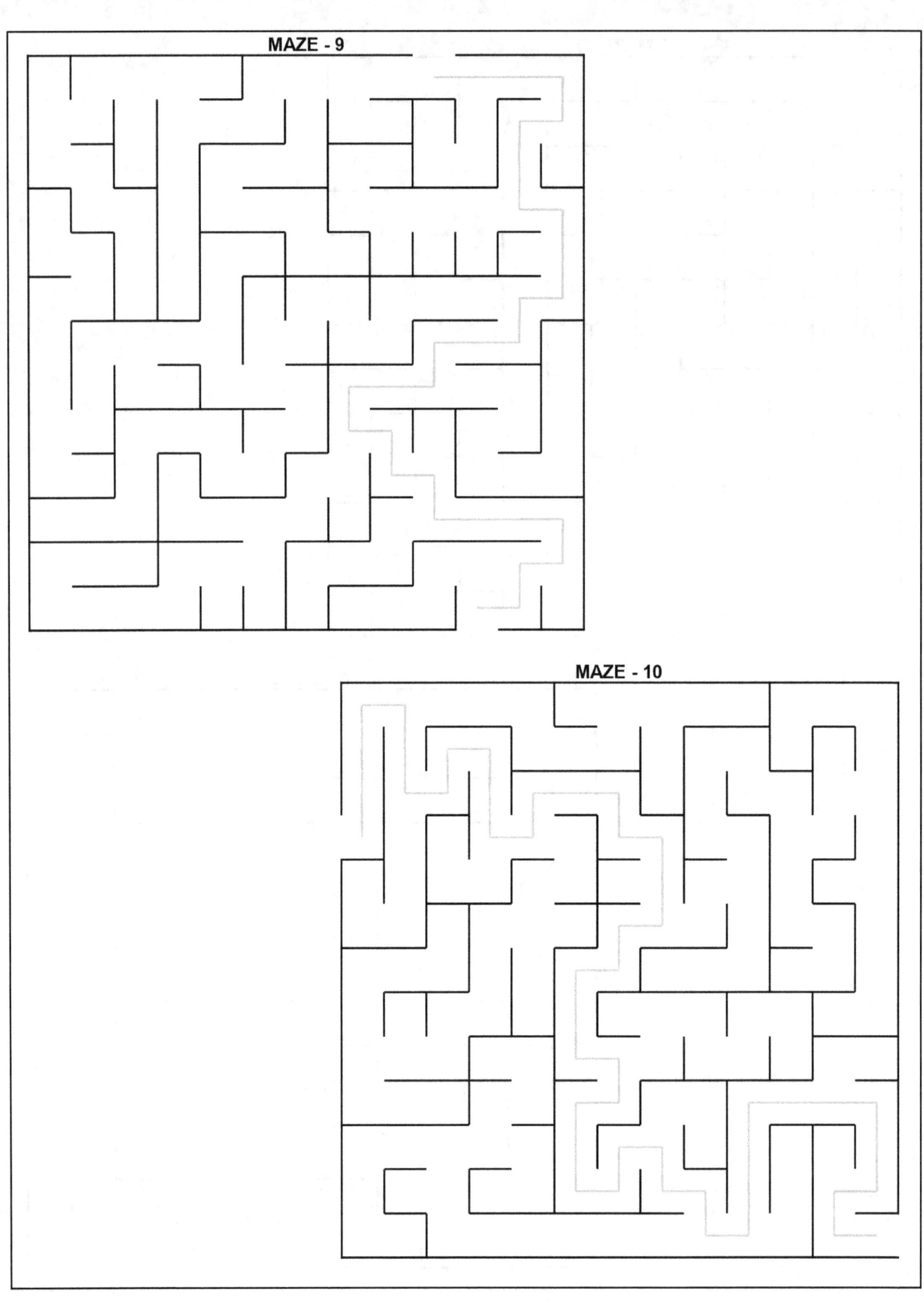

MAZE - 9

MAZE - 10

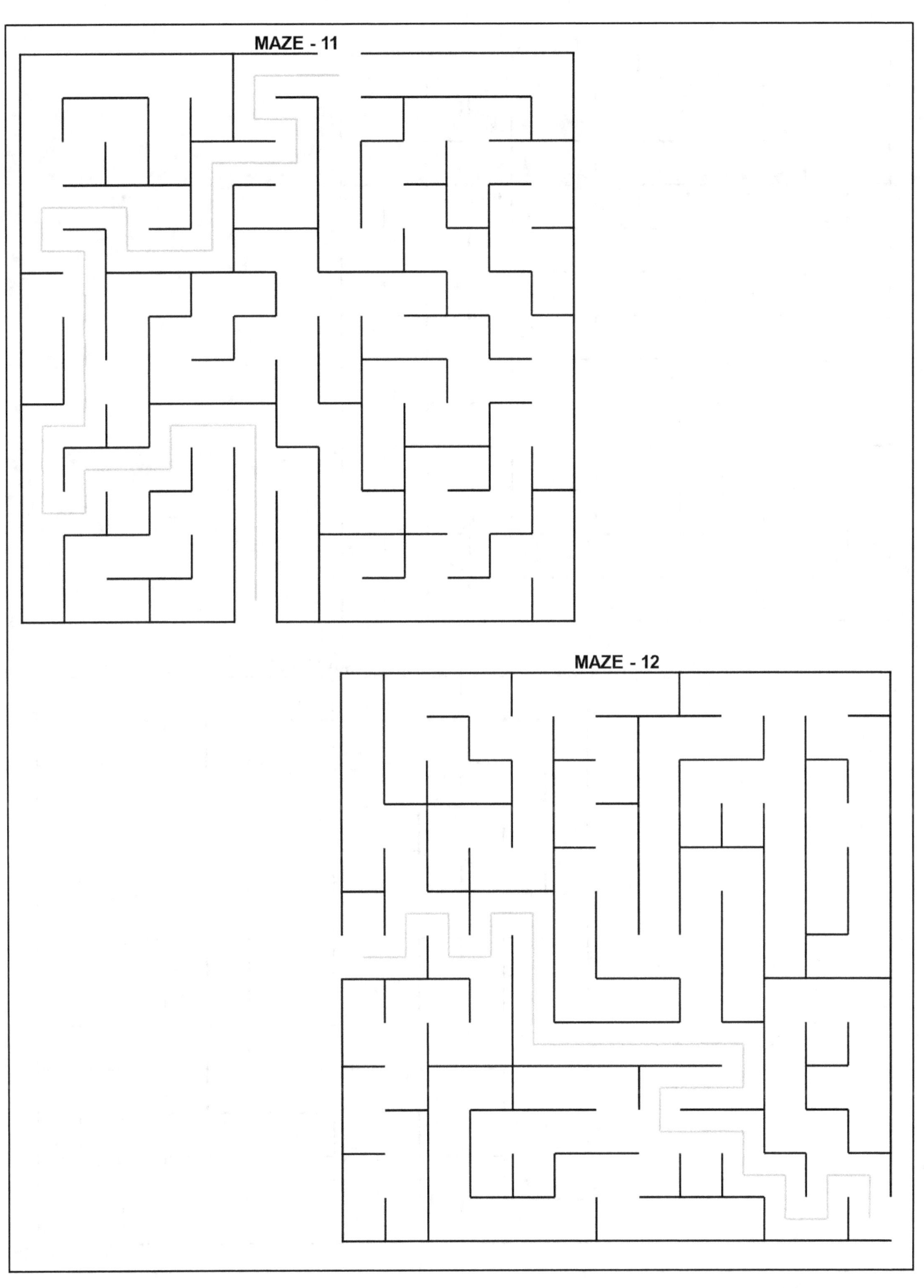

MAZE - 11

MAZE - 12

MAZE - 13

MAZE - 14

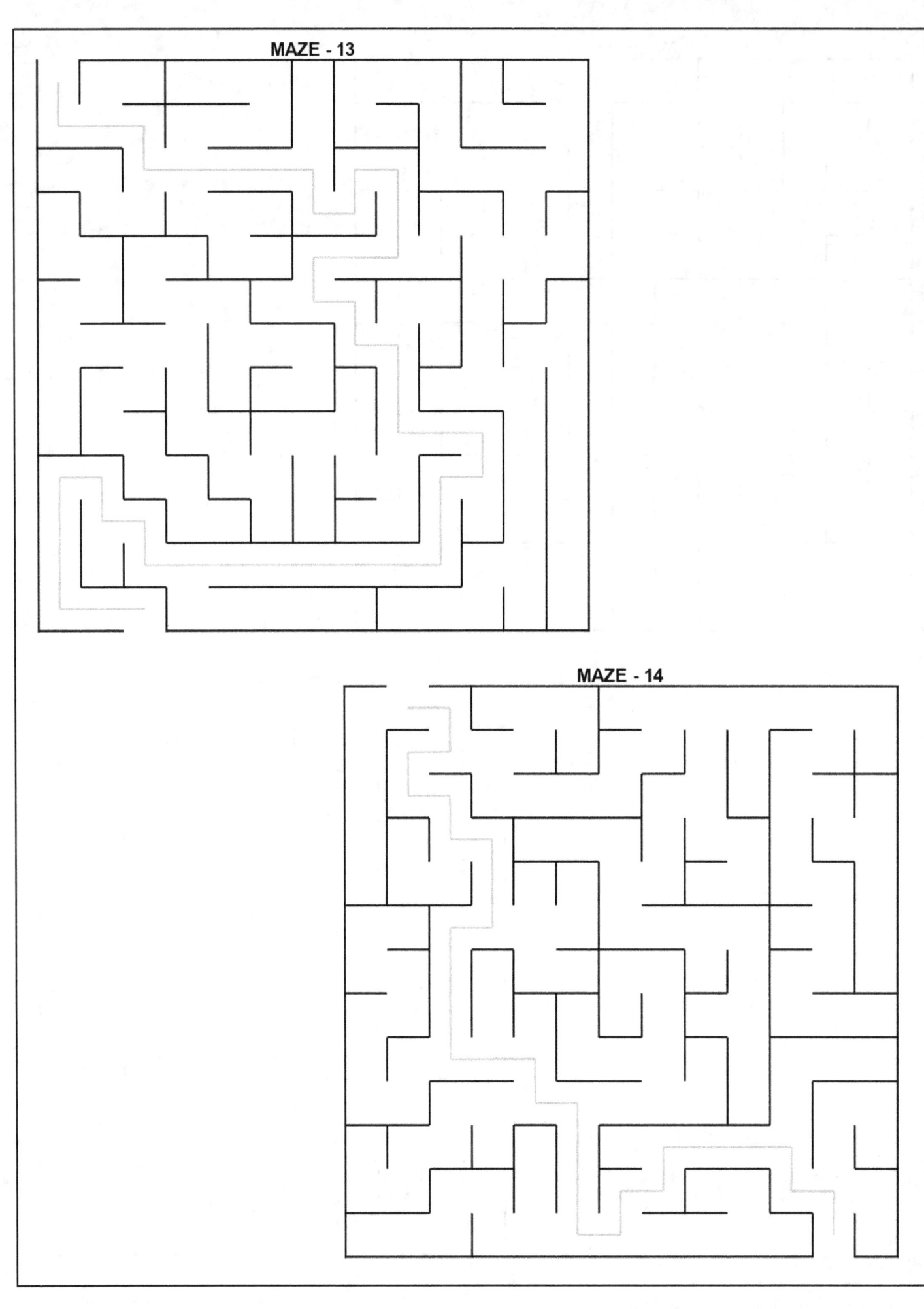

MAZE - 15

MAZE - 16

MAZE - 17

MAZE - 18

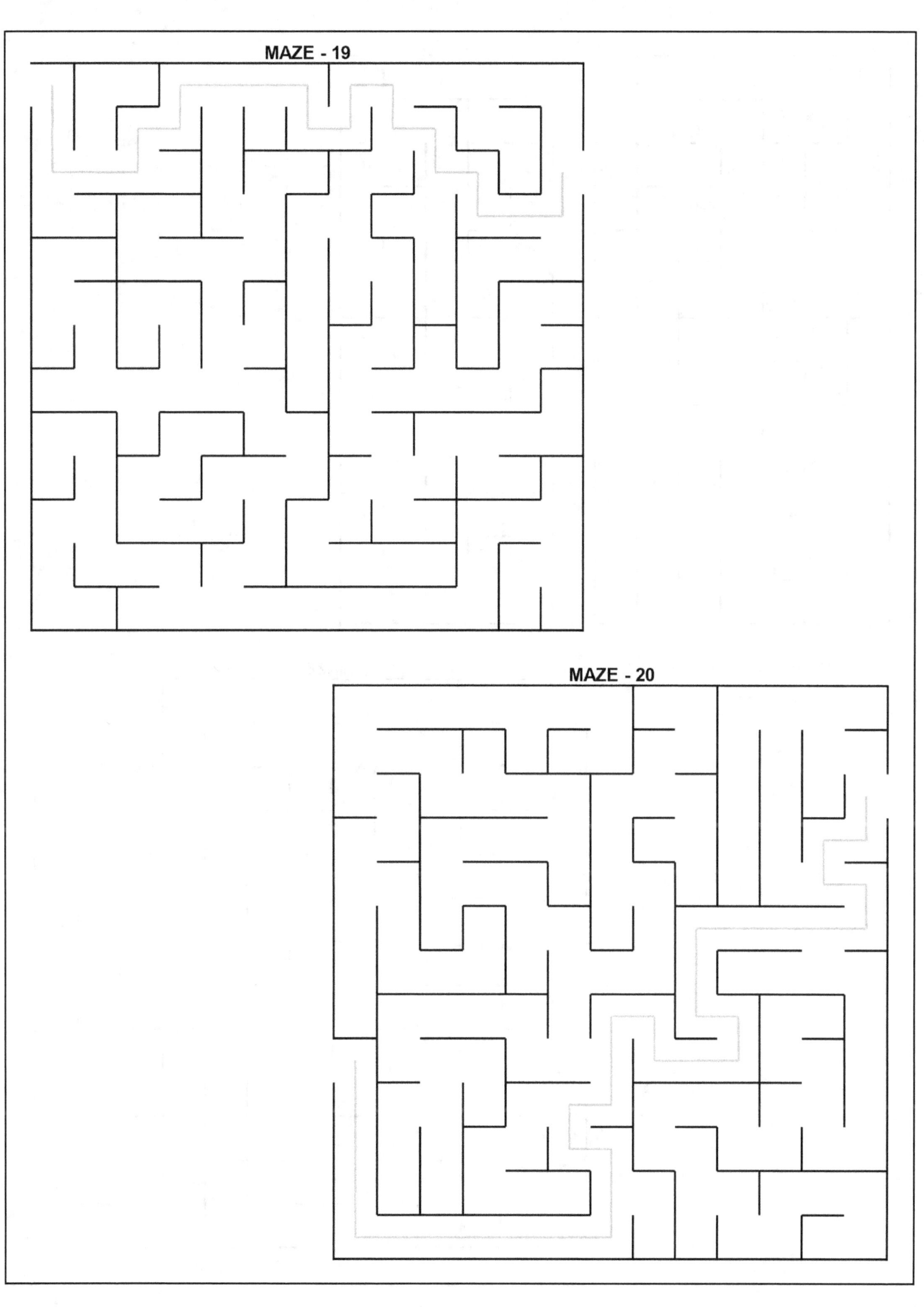

MAZE - 19

MAZE - 20

MAZE - 21

MAZE - 22

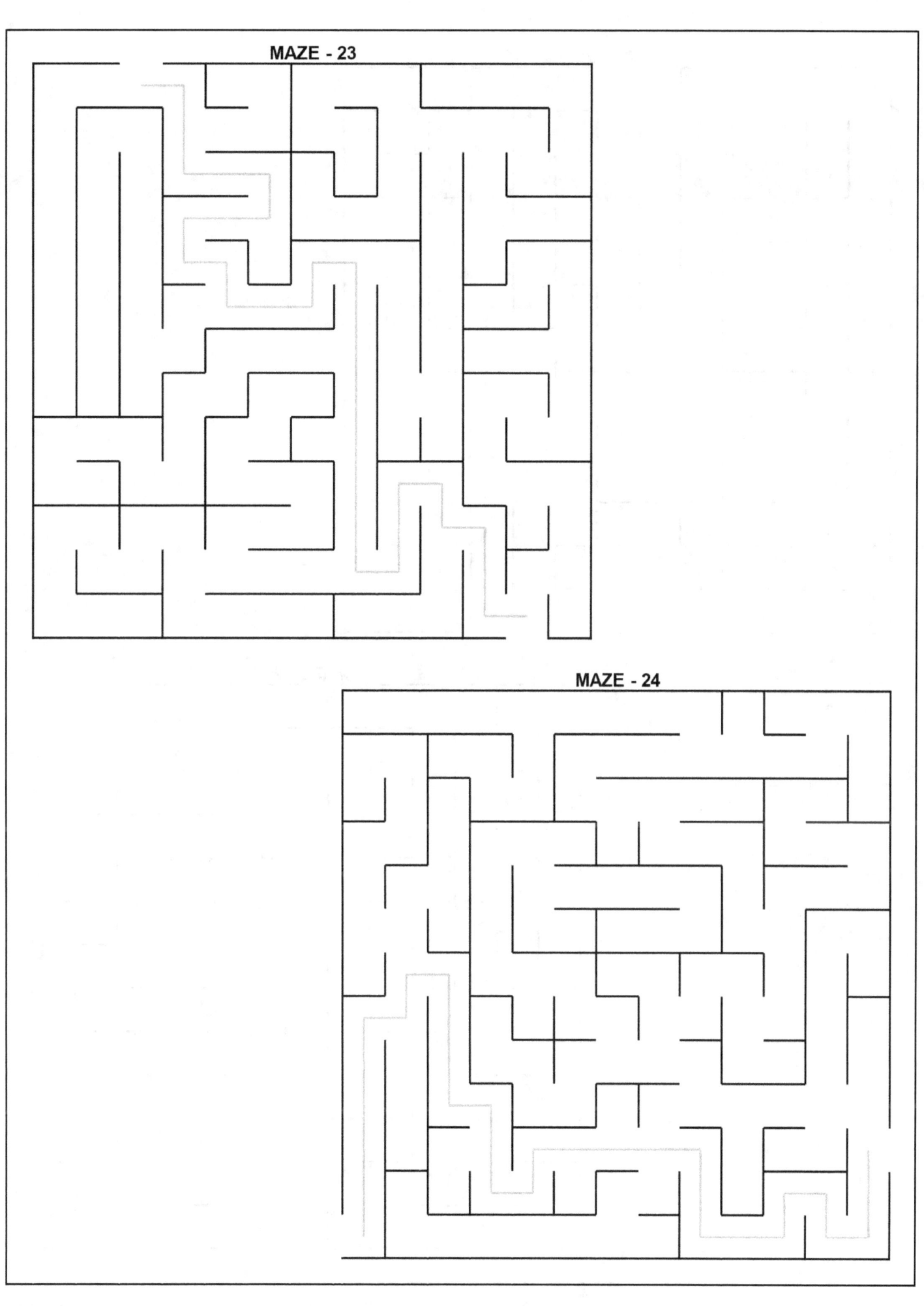

MAZE - 23

MAZE - 24

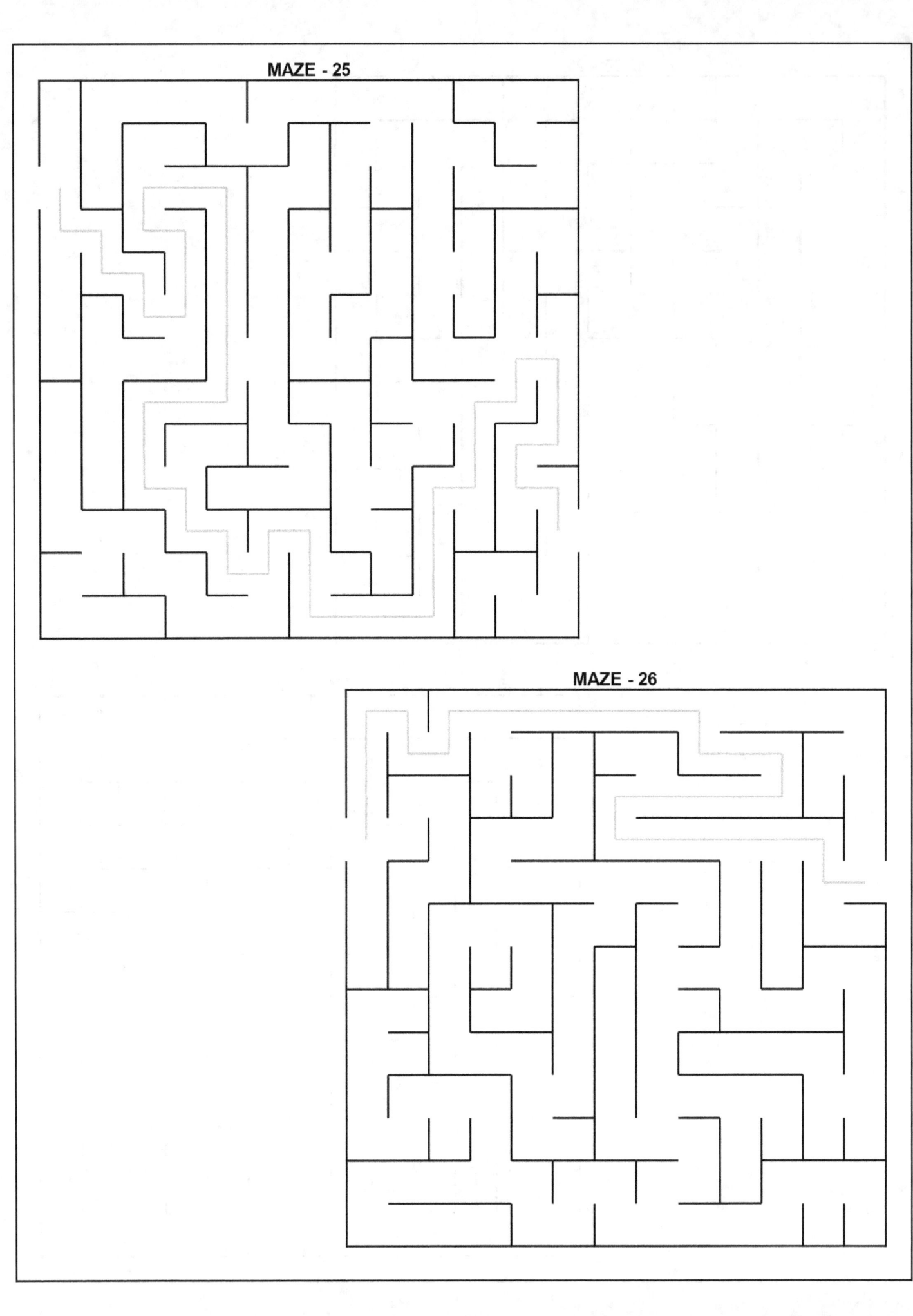

MAZE - 25

MAZE - 26

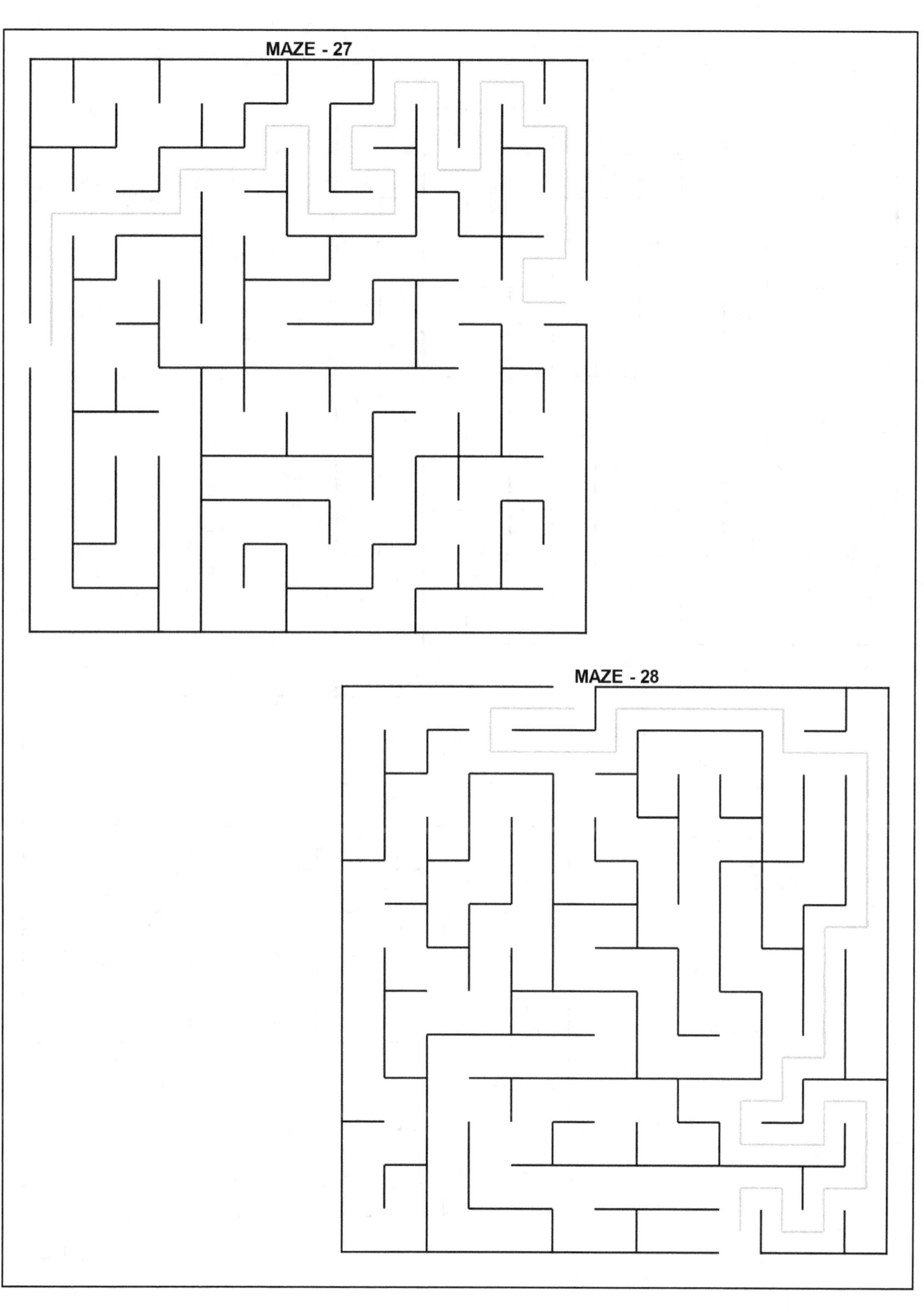

MAZE - 27

MAZE - 28

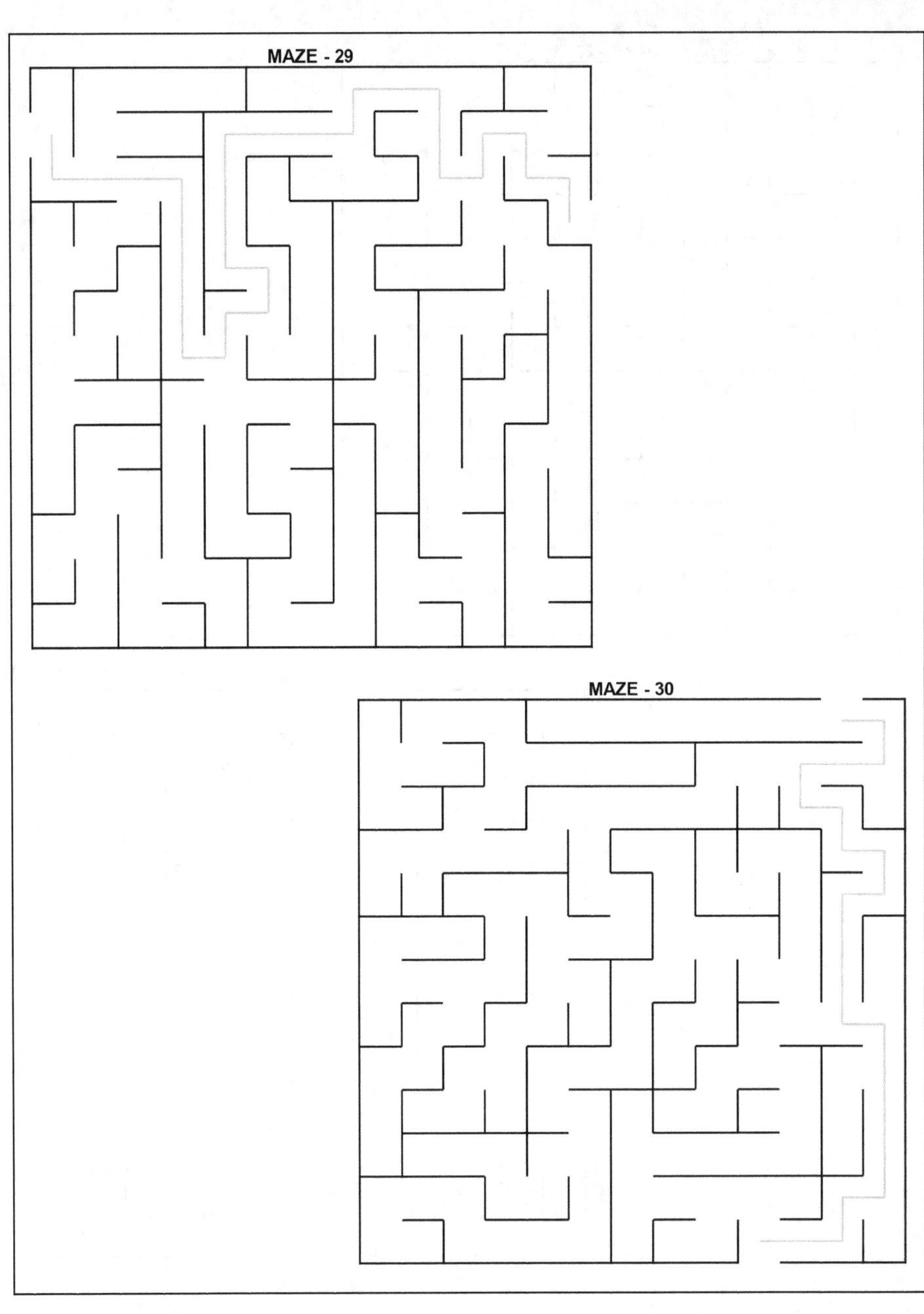

MAZE - 29

MAZE - 30

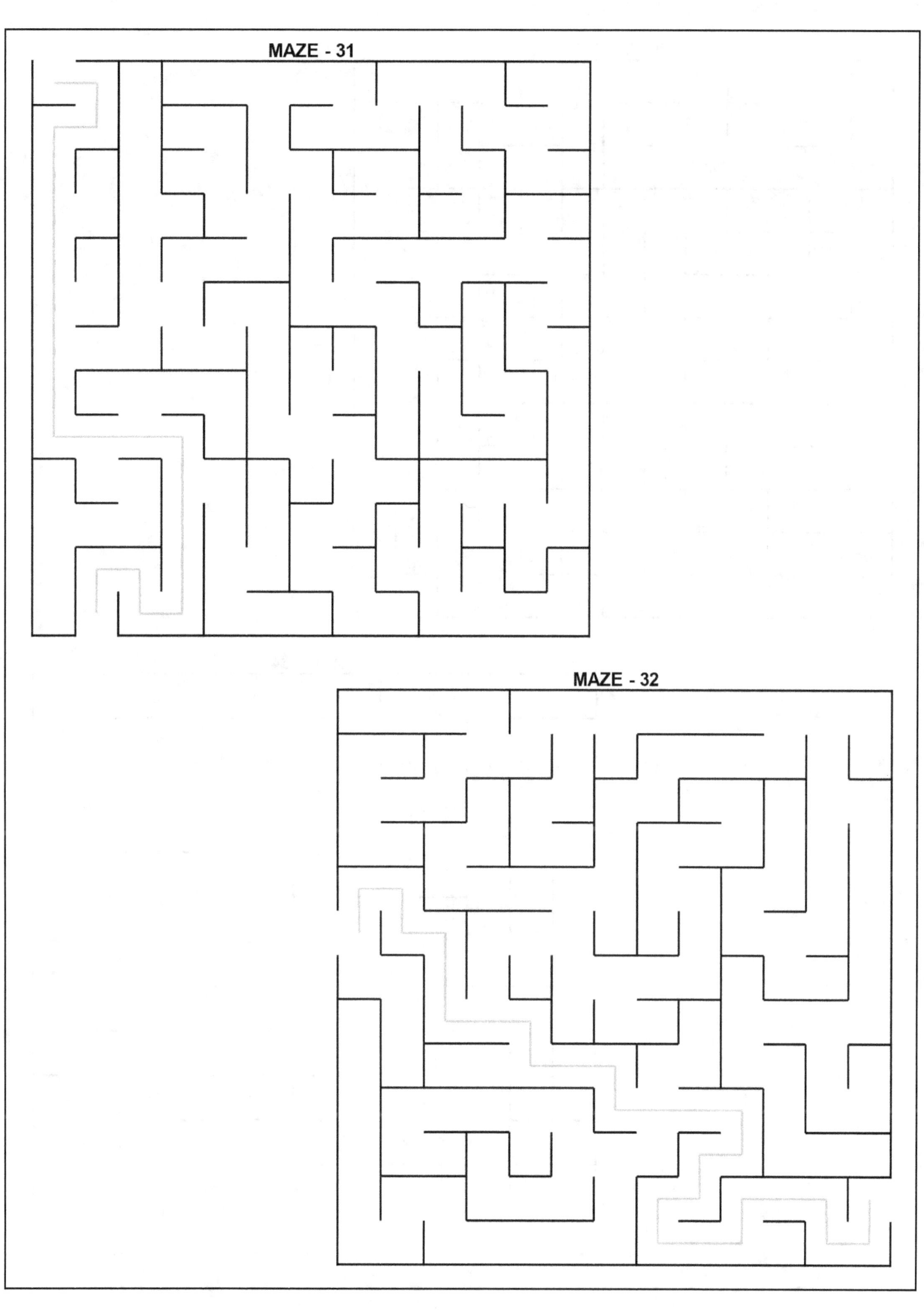

MAZE - 31

MAZE - 32

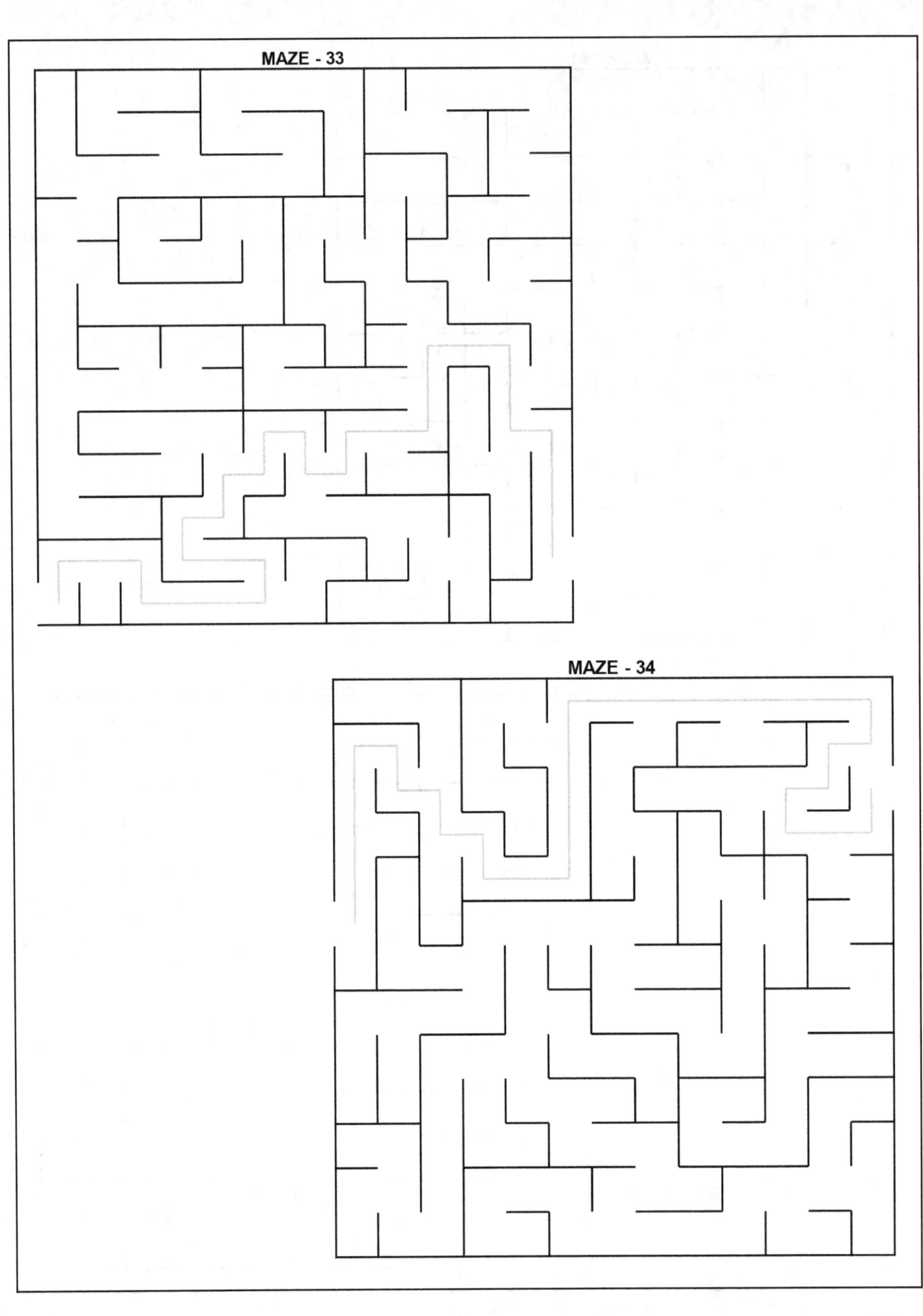

MAZE - 33

MAZE - 34

MAZE - 35

About the Author: Michael Eric Nelson is an American Author, Historian, Inventor, Mentor, Philanthropist, monetary Scientist and Numismatic Expert. His undergraduate degrees are in Business Administration and Economics, with minor concentrations in English, Geological Sciences and Military Strategy. He is a graduate of Norwich University, the Military College of Vermont. He has extensive international training from Brunel University, Buckinghamshire College of Business, London, England, UK. He has lived in Laguna Beach, California, Las Vegas, Nevada, Boston, Massachusetts, Northfield and Montpelier Vermont. He currently lives between Montpelier, Vermont and Rhode Island USA and Barcelona, Catalonia, Spain. He is a scholar of Spanish History, with concentrations on Spanish unification, as the Kingdom de España. Padrí Miquel el Americano Padrino invites you to take a journey with him through his eyes as he explores the Empire that forever changed the world for the better. Padrí Miquel sees that 100 years from now it shall matter not the size of his belly, or his bank account, but what may make the difference in the future is to be the change in the lives of our children.

Michael is Miquel in Catalá, the native tongue of the Catalan people, who occupy the city of Barcelona, which is the seat of government for the autonomous region of Spain, known as Catalonia. He is affectionately called Padrí Miquel. In Catalan Padrí is equal to the Castellano (español) nombre/nom (name) Padrino, which in ingles (english) is known as G-dFather.
www.G-dFather.com

Padrí Miquel writes both children's books and advanced programming / scientific works. The childrens books are a pass time hobby, written for his love for his g-d children which are many. He thrives knowing that the children will grow strong knowing they are loved and are armed with knowledge from the books he authors.

We hope you will continue to buy and read and give the books of Padrí Miquel to your little ones and know that every book you purchase, has all its profits (if any) going to philanthropic causes from animal rights, to people's rights in conformity with the United Nations UDHR (Universal Declaration of Human Rights). Padrí is an EagleScout (where the circle with the dot in the center logo is derived), inventor, author, mentor, monetary scientist, and numismatic expert.

We welcome your feedback and hope you enjoy the Novellas, Books and workbooks.

Visit: www.elPadri.com for more books, workbooks, poetry, activity books by Padrí Miquel.

www.ingramcontent.com/pod-product-compliance
Lightning Source LLC
Chambersburg PA
CBHW081450220526
45466CB00008B/2579